ARMY AVIATION IN VIETNAM

1961 ✧ 1963

An Illustrated History of Unit Insignia Aircraft Camouflage & Markings

RALPH B. YOUNG

VOLUME ONE

ILLUSTRATED BY STEVE DAVIS

FIGURE ARTWORK BY KEVIN LYLES

Cover Photograph and Frontispiece

93rd Transportation Company (Light Helicopter) H-21 helicopters, aircraft 'U' in trail, painted in the Army's standard, authorized high visibility scheme, with blue flight platoon markings, fly high above rice fields in I Corps along the South China Sea in mid 1962. This stunning picture taken by the legendary Dickey Chapelle, and published in the November 1962 issue of *National Geographic,* epitomizes Army aviation's early years in Vietnam. This photograph was chosen for the cover because of its juxtaposition of the often incredible beauty of Vietnam with the visually striking image of a helicopter at war.

U
ARMY

57
913
UTT
HELICOPTER COMPANY
SPECIAL 23D WARFARE
AVN DET
VIET NAM
3RD RRU
57TH
MED DET
HEL AMB

CONTENTS

DEDICATION

This book is dedicated to the memory of seven American patriots, who gave their lives in the cause of freedom, when, at 2200 hours, on 11 January 1963, their aircraft fell out of the sky, and with red flying lights whirling terrifyingly in the dark of night, spun into the ground on an island in the Mekong River.

CWO-2 Lawrence C. Hammond — 57th Transportation Company (Lt Hel)

CWO-2 Raymond C. Wilde — 57th Transportation Company (Lt Hel)

SP5 James D. McAndrew — 57th Transportation Company (Lt Hel)

PFC Boyce E. Lawson — 57th Transportation Company (Lt Hel)

Captain Donald B. Toth — 93rd Transportation Company (Lt Hel)

1st Lieutenant Lewis L. Stone — 93rd Transportation Company (Lt Hel)

1st Lieutenant Charles M. Fitts — 93rd Transportation Company (Lt Hel)

First published in the United States in 1999 by The Huey Company, Inc., P.O. Box 625, Ramsey, New Jersey 07446-0625.

Library of Congress Catalog Card Number: 99-73163

FIRST EDITION
Limited edition softbound; 500 numbered and signed hardbound copies.

ISBN - 0-9671980-1-1

Printer: Galvanic Printing & Plate Co., Inc., 50 Commercial Avenue, Moonachie, NJ 07074

Graphic Designer: Dennis Wheeler, Design First Inc., P.O. Box 371, Craryville, NY 12521

ACKNOWLEDGMENTS

A book is the culmination of many things, not the least of which is the help of others and the sacrifices of a few. The sacrifices of my family, particularly my wife Marlene, are legion and I am deeply indebted to her. I am also deeply indebted to four other people for any success of this book and for their considerable help and encouragement:

COL. Kenneth D. Mertel, USA Ret., who graciously consented to write the Foreword, is a skilled author, whose extensive writings about Army aviation in Vietnam have been one of the foundations upon which most subsequent histories of Army aviation's early years have been based. His articles titled "Vietnam Report" in Army Aviation Magazine were a constant source of inspiration and his unstinting giving of his time, energy, photographs and memorabilia have been truly appreciated.

The 1962-1963 "Army Aviator of the Year" COL. Emmett F. Knight, USA Ret. shared his own well written work, a limited edition scrapbook on the 57th Transportation Company (Light Helicopter), the 98th Transportation Detachment (CHFM) and the Vertol CH-21C Shawnee. His high profile position as Operations Officer with the 57th Transportation Company (Lt Hel) subjected Colonel Knight to my endless questions and his extraordinary memory of those early years and his expertise, helpfulness, and patience have been invaluable.

COL. Samuel G. Conley, Jr., USA Ret. whose boundless enthusiasm today mirrors the considerable talent, professional skill and enthusiasm that he brought to the command of the 611th Transportation Company (DS) in 1962-1963. Colonel Conley went to work, enlisting others, and giving unstintingly of his time to further research on Army aviation during the early days of the Vietnam War. The scope of his photographic documentation of Army aviation, both personalities and aircraft, in the 1962-1963 time frame is truly amazing.

CW4 Frank Baldwin, USA Ret., 93rd Transportation Company (Light Helicopter), who had a unique perspective, from the front seat of a Bird Dog, of the battlefield at Ap Bac on 2 January 1963. He willingly shared his knowledge, unique photographs and memorabilia from those early days and that significant battle.

The author would also like to thank the following people and organizations, for without their help this book could not have been written. My deepest appreciation to David E. Prewitt; COL. Robert L. Hoffman, USA Ret.; LTC. Joseph Henderson, USA Ret.; LTC. James E. Gray, Jr., USA Ret.; LTC. Milton P. Cherne, USA Ret.; COL. William C. Rudd, USA Ret.; LTC. Charles P. Callaway, USA Ret.; LTC. Leon W. Curry, USA Ret.; LTC. Robert F. Corneil, USA Ret.; Mel L. Brooke; John Cross; CW4 John F. Leonard, Jr., USA Ret.; CWO Marvin A. Farmer, Jr., USA Ret.; MAJ. George Tracy, USA Ret.; LTC. Leland F. Wilhelm, USA Ret.; CW4 Donald R. Joyce, USA Ret.; Wayne Mutza; James D. Sprinkle; Bob Chenoweth; Frank McGrane, TCAA; Dennis Buley; James L. Gilbert, Command Historian, INSCOM; Nguyen Trung Chau; Martin Windrow; Earl J. Dunn; Chris Miller, Cobra Company, who generously shared his knowledge and his vast patch collection; The National Archives, Suitland, MD; the Department of the Army Office of the Chief of Military History; the Department of the Army Institute of Heraldry; U.S. Army Transportation Museum, Fort Eustis; Army Aviation Association of America (AAAA); National Geographic Society; Nancy Palmer Agency; and Pictorial Parade Inc.

Finally I'd like to acknowledge my father, Raymond A. Young, Jr., who introduced me to a lifelong interest in aircraft and my mother, Lucille Rawley Young, whose love of history and art is reflected, although not necessarily in the way my mother would have envisioned, in this series.

LIEUTENANT COLONEL KENNETH D. MERTEL, Infantry

Major Ken Mertel had been ordered to Vietnam in late 1962 to assume command of the 18th Aviation Company (U-1A) headquartered in Nha Trang. Enroute he had received a promotion and, upon arrival at Tan Son Nhut Air Base was 'requisitioned' by Brigadier General Joseph W. Stilwell, Jr., son of old 'Vinegar Joe' Stilwell, who would not allow a Lieutenant Colonel, a grade too high, to command one of his aviation companies. Instead he became General Stilwell's Inspector General (IG), a combat IG as it were, with the mission to judge an aviation unit's morale, effectiveness in combat and the efficiency of support from higher headquarters, and report back to the General. He typically would spend three days in the field with a unit and often flew with their crews, as a co-pilot or as a door gunner, on many combat assaults. In addition he was tasked with countering U.S. Air Force allegations that the Army's armed helicopters were encroaching on the Air Force's close air support mission, as well as diplomatically diffusing some high command bias against the U.S. Army Special Forces.

In August of 1963 Lieutenant Colonel Ken Mertel assumed command of the 45th Transportation Battalion (Transportation Aircraft) at Tan Son Nhut. He was to be the last commander of the 45th Transportation Battalion as, on 24 September 1963, the Battalion was redesignated the 145th Aviation

Lieutenant Colonel Kenneth D. Mertel, Inf. Commander 45th Transportation Battalion (Transportation Aircraft), August 1963

Battalion with control of the 118th Aviation Company (AML) at Bien Hoa, the 120th Aviation Company (AML), UTT Helicopter Company and 145th Airlift Platoon at Tan Son Nhut. Lieutenant Colonel Ken Mertel then served as the first commander of the 145th Aviation Battalion until December 1963 when he returned to the United States.

Lieutenant Colonel Mertel was the subject of the following narrative, written by Col. Bernie Quedens, USA Ret. former Operations Officer of the 120th Aviation Company (AML). Lieutenant Colonel "Mertel was largely responsible for the more direct involvement of his headquarters in operational combat lifts by his subordinate units. But he also permitted the company commanders to run the show, while monitoring the assaults personally from a command and control helicopter, frequently flying in close proximity to the troop carrying helicopter formation. From time to time, he would be in radio contact with the commander or the mission controller during the execution phase of the mission, but he was never known to question the troop lift commanders judgment or interfere while the mission was in progress."

Colonel Mertel went on to serve two more tours in Vietnam, both with the 1st Cavalry Division (Airmobile). His first tour with the 1st Air Cavalry in 1965-66 was chronicled in his best selling, and still in print, book "Year of the Horse: Vietnam". During his second tour with the 1st Air Cavalry in 1970 he commanded the 11th Combat Aviation Group. His career combat decorations include the Silver Star, two awards of the Distinguished Flying Cross, fifty Air Medals, five Bronze Stars (one with 'V' device), four awards of the Legion of Merit, three Army Commendation Medals, and four awards of the Vietnamese Cross of Gallantry.

Colonel Kenneth D. Mertel is representative of the many named, and unnamed, professional Army aviators whose outstanding contributions and service, in the early years of the Vietnam War, are imprinted indelibly upon the story of Army aviation itself.

FOREWORD

Ralph Young has put together a most needed history of Army Aviation in Vietnam, beginning with this first volume entitled "Army Aviation in Vietnam 1961 -1963." He has captured the flying colors of Army combat aviation units and assembled data that had previously been stored away in a unit's history, in the memories of those who were there, or from their diaries and other writings. He skillfully weaves this all together in a superb fashion, making this a history not just of unit insignia and aircraft markings but also of a unit's challenges and accomplishments

Tribute is paid to the Transportation Companies' combat aviators, commanders and especially the crew chiefs, mechanics and other maintenance soldiers who "kept them flying" under extraordinarily difficult conditions, almost with "spit and chewing gum." Aircraft maintenance support and availability of spare parts was a major challenge, especially in those early days. This first volume is about the early Army aviation pioneers, who blazed the way. Had they failed, we probably would still be flying Bird Dogs or "Crop Dusters", which many, in and out of the Army, hoped would happen. But they did not fail.

Ralph tells the story of the disaster at Ap Bac, where we lost helicopters and crew, due primarily, to U.S. advisors not really knowing how to employ this revolutionary new weapon "airmobility" and the lack of training by the South Vietnamese units. He also covers the first use of armed helicopters in a so-called "Close Air Support Role" and how the Air Force did its best to stop the use of armed helicopters to provide fire support. Fortunately, they did not win, saner heads realizing in the Air Force, Navy and Marines that close air support was needed as well as fire support from the Army's armed helicopters.

Throughout these early pioneer years, Brigadier General Joe Stilwell, Jr. son of "Vinegar Joe" of World War II fame, guided all Army aviation units as Commander of the U.S. Army Support Group. He was a great supporter and friend of "Aviation Warriors." He became one himself by boot legging flight time in "my" Bird Dog with the Support Group Aviation Officer as his Instructor Pilot (IP). I ended up with the better of the trade since I got his twin engined L-23, which I though was safer to fly over the extensive jungles in Vietnam. After a suitable period of instruction, the General and his instructor landed at Soc Trang, the IP went to Operations to check something, and the General promptly took off for his first solo. He had to make a "go around" the first time but managed to safely put the bird on the ground on the second try. Some wondered if they would have to shoot him down, to get him on the ground! "Joe" saved many an aviator from unfair attacks by some of our "friends" who were still out to discredit Army aviators. "Joe" would not let that happen. Many thought him rough, profane and overbearing, however he always had the courage to fight for his "Aviation Warriors." None up the chain of command dared cross this "Warrior" when he thought someone was taking advantage of his aviators. He also became a door gunner on both the Huey and Flying Banana and flew dozens of combat missions. He was much revered, and respected, by his fellow aviators. General Stilwell died an airman's death, when he and a friend took off from the West Coast to fly to Hawaii in an old C-47 and disappeared at sea. It is believed that drums of extra fuel carried in the cargo hold came loose in rough weather, shifted, changing the center of gravity of the aircraft, causing them to go down in the ocean somewhere enroute. No one really knows. General Stilwell died "flying", his true love, next to "parachute jumping".

Ralph narrates the story of the first aviation battalion headquarters to come to Vietnam, the 45th Transportation Battalion, later redesignated the 145th Combat Aviation Battalion, units that I had the privilege to command. Later, with the arrival of more Army aviation units, the Battalion's responsibilities were split with the 52nd Combat Aviation Battalion at Pleiku and the Delta Aviation Battalion in Can Tho. He reminds us of the early days when we only had combat operations during the daytime (no night operations for those we supported.) At the end of the day, in Saigon, we would change to civilian clothes, go downtown in a pedi-cab or taxi to one of those superb French or Chinese Restaurants, all at most reasonable prices. How different it was in later years, on a second or third tour.

Ralph splendidly chronicles Army aviation in the early days in Vietnam. I know that both military and civilian readers will enjoy reading this book as much as I have and I look forward to the next volume.

Colonel Kenneth D. Mertel, Infantry, US Army Retired.
Williamsburg, Virginia, January 1999

AUTHOR'S PREFACE

This project, to document Army aviation unit insignia and markings, grew out of my early interests, hobbies and experiences. As I built innumerable models of P-51 Mustangs and Spitfires in the 1950's, I discovered that factual information on unit markings was either non-existent, hard to come by, plain wrong or total fiction; and this only ten years after the end of the war! Later, after college and Army ROTC, I went to Vietnam as a First Lieutenant, Transportation Corps, arriving during Tet of 1968 in Nha Trang, to join the Headquarters and Headquarters Company, 54th General Support Group, 1st Logistical Command. Part of my duties involved traveling throughout II Corps and I was amazed by the colorful unit artwork on Army helicopters and fixed wing aircraft. In the back of my mind was the scarcity of material on World War II unit insignia and believing that Army aviation unit markings would be subject to the same fate after the war, I determined to start recording what I could. While I started this project in Vietnam, I continued to research and document Army aviation units when I returned home to the United States in May 1969.

Many books, on various aspects of the Army's effort and the helicopter war in Vietnam, have been published, some extremely well researched, others full of errors and misinformation. None have yet illustrated the totality of the Army aviation experience in Vietnam, and it is my fervent hope that this volume will provide a start in correcting that deficiency. This is the first volume in a planned series intended to be the definitive, illustrated history of each and every Army aviation unit in Vietnam, down to the smallest aviation detachments and brigade aviation sections. It must be stated from the outset that this is not meant to be the official history of Army aviation in Vietnam, there are people far better qualified than I to write the official story. This illustrated history will briefly outline Army aviation, and an individual unit's, history in Vietnam, provide a chronology of units' arrivals and departures, discuss the painting, camouflage and marking of the Army's aircraft, and illustrate each unit's patch, insignia and how each unit marked their aircraft.

The author in Nha Trang in early 1969.

The 'personality' of a unit was most visibly expressed in the unit's 'unofficial' patches, insignia, call signs and aircraft markings. These 'unofficial' patches and insignia were the means a unit used to distinguish itself in combat or in a combat support role and the unit's chosen insignia represented the way a unit perceived itself and wanted to be seen by others. The present work began with the premise that the study of each and every unit's unique personality as expressed in its insignia, patches and aircraft artwork would collectively define Army aviation's Vietnam experience in a way that no other approach could ever hope to achieve.

Every effort was made to document by name those who made significant contributions to Army aviation history in Vietnam, although there are countless others, unnamed and sometimes unknown, whose contributions still shine brightly through these pages, commanders, pilots and crew chiefs, gunners and maintenance men, cooks and clerks, who kept the aircraft flying, made sure that the unit's mission was accomplished, and sometimes, gave their lives. However, having said all that, the author takes full and sole responsibility for any errors, omissions, oversights and just plain mistakes that this present volume may contain.

Ralph B. Young, New Jersey, January 1999

INTRODUCTION

To better understand the geographical locations, topography, make up of the population, religion and the form of government of South Vietnam, as mentioned in the text, a brief description of the Republic of Vietnam as it was in the early 1960's is necessary.

TOPOGRAPHY

South Vietnam, known in the 1960's as the Republic of Vietnam, had an area of 65,948 square miles, and was situated on a narrow strip of land bound by the South China Sea to the east, the Kingdoms of Laos and Cambodia to the west, the Gulf of Siam to the south and the communist Democratic Republic of Vietnam to the north, across the 17th parallel. The topography of the country is divided into four main regions, the lower third of the country is dominated by the Mekong River Delta, a frequently marshy estuary of the Mekong River system. This broad, flat, low-lying river delta region was a rich agriculture area, the 'rice bowl' of the south. To the north and east of Saigon were the eastern provinces of South Vietnam, areas of low-lying tropical rain forest, upland forest, and the southern end of the rugged Annamite Mountain Chain. Central Vietnam is divided into a narrow coastal strip, with intense rice farming, and a broad inland plateau area, the two separated by the Annamite Mountains.

Saigon and the Mekong Delta region experienced a tropical climate year-round, while the central highlands were generally cool from October to March, the temperatures sometimes falling to around 50° F.

POPULATION

The population of South Vietnam in 1962 was approximately 14.9 million, 90 percent of whom were ethnic Vietnamese, the rest composed of minorities such as the Chinese, Cambodians, Chams and several distinct ethnic minorities, known as the Montagnards, or mountain tribesmen. Montagnards were divided into many tribes such as the Rhade, who lived near Ban Me Thuot in the Central Highlands. These tribes lived primitive, semi-nomadic lives in the mountains and high plateau of Central Vietnam.

Saigon, the Capital City of the Republic of Vietnam with a population of 1.4 million in 1962, was located on the Saigon River, a navigable waterway. Saigon was also served by Tan Son Nhut airfield, a dual use civilian and military air base which was approximately ten miles to the west of the downtown area, a short taxi ride away.

RELIGIONS

The religion of most Vietnamese was a mixture of Taoism and Buddhism, both derived from ancient Chinese and Indian beliefs. There were also approximately 1.5 million Roman Catholics in Vietnam, and the Montagnards practiced Animism. Buddhism was perhaps the most visible of Vietnamese religious beliefs, in both elaborate temples, pagodas and shrines and in the saffron-robed, clean-shaven monks who seemed to wander tirelessly about, fed and supported by the people. In Vietnam, Buddhism's influence was pervasive in the arts, philosophy and literature, the theater, the people's custom's, myths and legends.

During the early 1960's Buddhist leaders were in open conflict with President Diem, who was a Roman Catholic. The Buddhists wanted an end to the government favoritism being shown to Catholics, and an end to Diem's repression of all political and religious opposition. Because of Diem's fear of coup attempts, his authoritarian solution was to suppress any form of dissent, including flying of the five-colored Buddhist flag in public. Tensions had escalated in May when a street demonstration in Hue was confronted by government troops, and demonstrators were killed. The Buddhists fought back, on the world stage, for on 11 June 1963 the first Buddhist monk, Thich Quang Duc, immolated himself in public. Others followed in what became a Buddhist uprising against the government.

The growing Buddhist opposition to Diem's government was one of the factor's that turned the Kennedy administration against

Figure 1. Topographical Map of South Vietnam

Diem in mid 1963 and tacitly sanctioned the coup which would result in Diem's overthrow, and death, on 2 November 1963.

Catholicism in Vietnam dated back to the 16th century and was one of the major religions in South Vietnam claiming approximately 11 percent of the population. The generally higher educational levels of Catholics tended to place them in positions of influence within the government and society. Over half a million Catholics migrated from the north and were settled in South Vietnam when the country was separated after the Geneva Accords of 1954.

One group, in particular, deserves special mention, the 300 Chinese Catholics of the Reverend Augustine Nguyen Lac Hoa who settled in Binh Hung in the Delta. Father Hoa was a refugee from China and a former Colonel in the Nationalist Chinese Army. No roads led to Binh Hung, and the Viet Cong controlled the surrounding countryside. Once in Binh Hung Father Hoa had organized his own self-defense force, which he named "Sea Swallows." The Sea Swallows quickly made a name for themselves as a well-disciplined and courageous fighting force, which attracted overt and covert U.S. aid. By 1962 they were a valuable ally in the fight against the Viet Cong, Binh Hung becoming known as the "village that refuses to die", and was regularly visited by helicopters and crewmen of the 93rd Transportation Company (Lt Hel) as the safest means to transport supplies was by air. The Sea Swallows, for example, were given the new AR-15 Armalite rifle, forerunner to the M-16, even before the U.S. Army.

GOVERNMENT

After the 1954 Geneva Accords separated Vietnam into two political entities, the people of South Vietnam, in 1955, chose a republican form of government in a national referendum, electing as President Ngo Dinh Diem. This republican form of government was organized under a constitution, of 26 October 1956, which provided for a strong executive office, that of President, with a five year term, a single chamber legislative assembly of 123 members, each for three year terms, and an independent judiciary. At first President Diem, with the support of the U.S. but not the French, set about consolidating his government's position and moved swiftly to crush the powerful private armies that roamed the countryside and were even entrenched in Saigon. Using a combination of incentives and force, President Diem succeeded in ousting, from Saigon, the Binh Xuyen, a band of former river pirates, and subdued the quasi-political religious armies of the Hoa-Hoa and the Cao Dai. Action against former Viet Minh members was also initiated, so that by 1958 the Saigon government was in undisputed control of South Vietnam. It was in response to the success of Diem's tactics that the communist North Vietnamese, in 1959, formed the National Liberation Front (NLF) and began to infiltrate trained cadre back into South Vietnam. On 18 January 1962 Radio Hanoi seemed to drop all pretense of non-involvement in South Vietnamese affairs when it announced the formation of the Marxist-Leninist Vietnamese People's Revolutionary Party to help the National Front for the Liberation of South Vietnam.

MAP OF SOUTH VIETNAM
August 1962

Nha Trang AUTONOMOUS MUNICIPALITY

ARMY AVIATION UNITS

Army Republic of Vietnam Infantry Divisions and assigned Areas of Operations

Figure 2: Map illustrating the boundaries of I, II, and III Corps, South Vietnam, in August 1962 and the Areas of Operations and locations of the ARVN Divisions. The basing of the Army aviation units, a battalion headquarters, three Army cargo helicopter companies, a fixed wing unit, a direct support maintenance company and the helicopter ambulance detachment, in relationship to the Corps areas, and the ARVN Divisions, is apparent. The IV Corps area was not created until the end of 1962.

SECTION I • War in Indochina 1946-1961

After the surrender of Japanese forces at the end of World War Two, France prepared to reassume control of her former colonies in Indochina. A collaborationist French presence had been maintained throughout the Japanese occupation while various nationalistic and communistic Vietnamese factions had joined, under the Communist Revolutionary Ho Chi Minh, in resistance to the Japanese. Allied victory brought some measure of national authenticity and international respect to the resistance, which had been assisted in their war against the Japanese by covert American aid. The French government was obliged to hold talks with Ho Chi Minh's Vietnam Independency League, the Viet Nam Doc Lap known as the Viet Minh. Assured initially of considerable autonomy within a French Union, Ho Chi Minh's position deteriorated subsequently due to political instability in France and powerful colonial groups which were not willing to relinquish either their authority or their vested interests by allowing any autonomy to the Viet Minh. As the French government's position hardened and their military strength grew so did a confidence that the French could impose their will on any organized resistance, and it became inevitable that incidents between the Viet Minh and French forces led to renewed conflict in Indochina in 1946, in what became known as the First Indochina War.

From the beginning the French military consolidated their hold on urban centers, transportation and communication networks, while maintaining static defensive lines and positions. In addition the French recruited and trained local units to augment their forces. These steps were not sufficient to assure victory and the war dragged on, neither the French nor the Viet Minh being able to decisively defeat the other. France's protracted war attracted notice in the United States, where, it had been concluded, free nations were inextricably bound up with America's own national security interests. In spite of increased U.S. military aid, the French government rightly concluded that other steps would have to be taken if the conflict was to be resolved in their favor. Belatedly, in February 1950, the French government granted Vietnam, Laos and Cambodia, known as the Associated States of Indochina, autonomy within the framework of a French Union, the autonomy which had been promised, and then withheld, in 1946.

Supporting a nation's right to independence and self-government, United States government policy had not allowed any direct assistance to Vietnam, Laos or Cambodia as long as the French ruled, but France's granting of autonomy prompted the U.S. government to recognize these states and provide them with U.S. military aid. Accordingly, in May 1950, the U.S. Secretary of State Dean Acheson announced that a small military assistance advisory group would be established to provide military equipment to the Associated States of Indochina and the French. This small detachment, known as the Military Assistance Advisory Group (MAAG), Indochina, was in place in Saigon by the end of 1950, commanded by Brigadier General Francis G. Brink. While they accepted U.S. military aid, the French retained responsibility for the conduct of the war and training of French Union forces, such as the Vietnamese National Army. Kept at arm's length by the French, MAAG, Indochina had little responsibility and no authority. In 1952 Major General Thomas J.H.

Trapnell succeeded General Brink as head of MAAG, Indochina. The French war efforts were stalled when, in May 1953, General Henri-Eugene Navarre assumed command of the French Union forces. General Navarre recognized that the static defense posture of the French Union forces was not only ineffective, but had ceded the initiative in the war to the Viet Minh. In a change of strategy, known as the Navarre Plan, General Navarre increased the size of the Vietnamese National Army to replace French forces in their defensive positions. As French forces became available, they were then to be used in mobile units whose task would be to locate the Viet Minh in the countryside and force them to stand while superior firepower would destroy them. The larger Vietnamese National Army forces, U.S. military aid along with reinforcements from France would return the military initiative to the French, as projected by the Navarre Plan, by 1955.

The war had never been popular at home in France and pressure was put on General Navarre, from the French government, to bolster public opinion with some good news from the battlefield. General Navarre was to initiated his plan prior to having all its elements in place, when, late in 1953 he ordered a small outpost fortified in a valley in northern Vietnam, near the border with Laos, named Dien Bien Phu. He intended to both block any Viet Minh advance into Laos as well as to draw the Viet Minh into a set piece battle where he could use superior French firepower to defeat them. In what was both a tactical and strategic surprise for the French, Viet Minh General Vo Nguyen Giap accepted the challenge, laid siege to the fortress and began to amass artillery and anti-aircraft guns in the hills around Dien Bien Phu. When ready the Viet Minh used their artillery to systematically bombard the French garrison and anti-aircraft guns to deny the French the ability to supply it by air, while laborers dug trench lines around the fortress, from which the Viet Minh troops could assault the French positions. By 8 May 1954 the battle was over and the French garrison of Dien Bien Phu was in Viet Minh captivity.

This unanticipated defeat at Dien Bien Phu stunned both the French government and the French public, and doomed any further French support for the war in Indochina. The French government hastily agreed to end the war, and on 20 July 1954 the recognition of the emergence of an independent Vietnam, Cambodia and Laos was formalized in the Geneva Accords. While France had all but conceded the North to Ho Chi Minh, in the South French strength, influence and protégés, such as Emperor Bao Dai, made it possible to create an independent state. The U.S., as a condition for participation in the talks in Geneva, Switzerland, had insisted that the southern half of Vietnam remain separate and this was accomplished by dividing the country in two at the 17th parallel. However the Geneva Accords settlement provided for the eventual reunification, through elections, of both South and North Vietnam politically, but recognized that until that time occurred, provisions were to be made to allow those Vietnamese who found themselves on the wrong side of the 17th parallel to relocate to the side of their choice.

The success of North Atlantic Treaty Organization in Europe prompted the United States to prepare the groundwork for a similar organization in Southeast Asia. Named the South East Asia Treaty Organization (SEATO) this treaty was intended to prevent further communist advances, the main United States concern being an overt invasion of South Vietnam by the North. In September 1954, the SEATO Treaty provided for, amongst other things, the active defense of South Vietnam against a communist invasion.

Still determined to retain as much presence as possible in South Vietnam, France correctly saw that a growing American interest was a threat to their continued, if subdued, influence. In April 1954 MG John W. O'Daniel had replaced General Trapnell as the third U.S. Army commander of MAAG, Indochina. In October 1954 an Aviation Division within MAAG, Indochina was established supported by an Army aviation flight detachment. This newly created MAAG Flight Detachment was the first Army aviation unit, with six officers and four aircraft, to be assigned to Vietnam.[1]

Meanwhile the Head of State and French protégé, Emperor Bao Dai, better known for his playboy lifestyle than political savvy, had become a virtual figurehead. To resolve the leadership vacuum, Emperor Bao Dai asked a relatively unknown Catholic, Ngo Dinh Diem, to be the Premier of South Vietnam. Diem had the tacit approval of the U.S. but not of the French, and this lack of support weakened the French position, occurring, as it did, simultaneously with the withdrawal of their military forces. Bao Dai lost what little power base he had and Ngo Dinh Diem called for a National referendum in which he was able to oust Bao Dai and in October 1955 become President of the newly proclaimed Republic of Vietnam, which established Saigon as its Capital. The scheduled conference between South and North Vietnam, proposed in the Accords for 1956, whose ultimate aim was the uniting of the two Vietnams through elections, was rejected by the new government. Inevitably this led to a renewed political and military struggle for South Vietnam.

MAAG, Indochina was redesignated the Military Assistance Advisory Group, Vietnam on 1 November 1955. As the French disengaged, severe problems became apparent as the French military had not trained the Vietnamese to handle logistical operations. This logistical nightmare was inherited by Lieutenant General Samuel T. Williams when he succeeded General O'Daniel as commander of MAAG, Vietnam on 18 November 1955.

Recent U.S. experience in Korea, and concerns in the United States about a cross border invasion, contributed to the decision to train South Vietnamese forces to contain a conventional North Vietnamese Army (NVA) invasion across the 17th parallel. The Democratic Republic of Vietnam, North Vietnam, was not preparing to storm across the Demilitarized Zone (DMZ), at least not in 1955, but was planning to subvert and infiltrate the south. The northern communists trained and returned south many of the former Viet Minh cadre who had relocated north after the partition. It was upon this foundation that they determined to build the revolution that would topple the South. In 1959 officers and noncommissioned officers were infiltrated from the north into the south and the struggle was joined in earnest. The French had trained the Vietnamese Army to replace its own forces in static and prepared defensive positions, and to hold the urban centers, transportation and communication networks, the very same strategic defense concept upon which the Viet Minh had honed their combat skills. It was no surprise that the communists early in the conflict showed a quick grasp of South Vietnamese military weaknesses, and, even though armed with often primitive or obsolete weapons, they accounted well for themselves. By the end of the year the communists had gained the initiative in the South.

Throughout 1960 communist guerrillas stepped up their activity, the South Vietnamese military not being equipped by training or tactics to effectively counter this type of warfare. President Diem, who feared a military coup, did not promote officers that he did not trust, and this led to a docile military leadership ill-suited by temperament to aggressively pursue the war. The Army Republic of Vietnam (ARVN) became a virtual do-nothing army led by politically compliant officers, while the communists took control of the countryside, becoming particularly well entrenched in the Mekong Delta region. In December 1960, the National Liberation Front (NLF) for South Vietnam was officially founded, representing the political arm of the insurgency. The NLF attracted some nationalists to its ranks but, in reality, it was organized and directed from the North. While the political arm was formally known as the NLF, this organization and its military arm were more commonly labeled the Viet Cong (VC), or Vietnamese Communists, by the South Vietnamese and U.S. forces.

The month before, in November, President Kennedy had been elected to office in the United States. The Kennedy administration viewed the recent enemy political and military initiatives with mounting concern. Senior U.S. officials and advisors in South Vietnam, known as the 'Country Team', prepared a planning document outlining the military, political, economic and psychological requirements necessary to defeat the communist insurgency. General Williams was succeeded by Lieutenant General Lionel C. McGarr at MAAG, Vietnam on 1 September 1960 as the Country Team's plan to defeat the insurgency began to be circulated in Washington. The Kennedy administration were favorably impressed by this plan and decided to increase funding and military aid to Diem in South Vietnam. In April 1961 President Diem easily won re-election, and so in May Vice President Lyndon B. Johnson visited South Vietnam to both confer with President Diem and to show U.S. resolve to both the North and South Vietnamese. The Vice President's trip was also intended to sound out Diem. His reply to the Vice President was adamant, Diem rejected allowing U.S. combat troops into the country, but he left an opening for an increase of U.S. forces to conduct training.

Then the communists launched a major offensive in central Vietnam, and significantly, on 18 September 1961, the Viet Cong captured the provincial capital of Phuoc Vinh, in Phuoc Long Province, which was only sixty miles north of Saigon. The enemy took the town early in the morning and held it all day long, publicly beheading the province chief in front of a huge crowd. A second major Viet Cong attack, on 22 September in Darlac Province, unnerved President Diem who reconsidered his opposition to U.S. forces and began to press for a bilateral defense treaty with the United States. By October 1961 President Diem was willing to accept U.S. combat troops.

To get a first hand report on the military situation, President Kennedy sent General Maxwell D. Taylor to Vietnam. The question General Taylor had to consider was not whether to intervene militarily but how it was to be done. General Taylor found that there was a crisis of confidence in the south, that the ARVN was on the defensive and that the current tactics being employed by the Diem government would not defeat the communists. General Taylor concluded that the ARVN could immediately benefit from increased mobility. When he returned to Washington General Taylor submitted these recommendations in his report - the "introduction, under MAAG operational control, of three helicopter squadrons - one for each corps area - and the provision of more light aircraft, as the need may be established..." [2] Dean Rusk and Robert McNamara, the Secretaries of State and Defense, further recommended to President Kennedy the introduction "as speedily as possible" of "units of modest size required for the direct support of South Vietnamese military effort, such as communications, helicopter and other forms of airlift, reconnaissance aircraft, naval patrols, intelligence units, etc." [3]

This was the situation on 1 November 1961 as alert notifications for deployment were sent out to two H-21 Transportation Companies (Light Helicopter). Set in motion were events which would lead to the coming of age of Army aviation.

HEADQUARTERS FORT LEWIS
Fort Lewis, Washington

AMNLE-GCT-P&O 8 November 1961

SUBJECT: Movement Order Number 3
Movement of 57th Transportation Company (Lt Hel)(H21) and 98th TC Detachment (CHFM) from Fort Lewis, Washington to Overseas Exercise Area

THRU: Commanding General
4th Infantry Division
Fort Lewis, Washington

TO: Commanders Indicated in Distribution

1. Prepare for foreign service and move personnel and equipment of the following units from Fort Lewis, Washington to Overseas Exercise Area through Oakland Army Terminal, TCS:

UNIT	TOE	SHIPMENT NR.	AUTH STRENGTH
57th Trans Co (Lt Hel)(H-21)	55-57D	2080-A	147
98th Trans Det (CHFM)	55-500R	2080-B	56

2. Movement Instructions. a. This is a temporary Change of Station (TCS) for period in excess of 30 days. Detachment remains as presently assigned until embarkation, at which time it will be attached to Overseas Commander for period of TDY.

b. Provisions of Chapter 3, less paragraph 6b, AR 220-10 (POM) apply.

c. Provisions of AR 612-35, as changed, do NOT apply in the selection of personnel for participation except that immunization requirements will be met and the following personnel will NOT participate:

(1) Personnel with an ETS prior to 30 May 1962.

FOR OFFICIAL USE ONLY

Figure 3. Copy of the Movement Orders, dated 8 November 1961, which would send the 57th Transportation Company (Lt Hel) (H-21) from Ft. Lewis, Washington to Saigon, Vietnam.
Courtesy of Col. Emmett F. Knight, USA Ret.

Right: **A portrait of the ARVN. At times a fine, capable soldier particularly when well led. At other times, if poorly led, he could be corrupt and ill-disciplined.**
Photograph courtesy of Col. Sam Conley, Jr., USA Ret.

Far Right: **Viet Cong captives under guard in the early 1960's. Referred to as 'raggedy-assed little bastards' by the American high command, Army aviators often came to respect this elusive enemy.**
Photograph courtesy of Col. Sam Conley, Jr., USA Ret.

Shortly after the October 1961 relocation of the 18th Airfield Operating Detachment (AOD), with two U-IA Otters, from Okinawa to Soc Trang, Vietnam, two Army cargo helicopter units received alert notifications to prepare for deployment to an undisclosed Overseas Exercise Area. After being ferried to the Oakland Army Terminal, in California, the H-21's of the 8th Transportation Company (Light Helicopter) (H-21) and the 57th Transportation Company (Lt Hel) were cocooned, to prevent salt water damage, and loaded onto the flight deck of the USNS Core. Both unit's assigned H-13's were boxed and shipped separately aboard a conventional cargo ship along with the unit's vehicles and other equipment and stores. On 21 November 1961 the USNS Core departed the west coast, the ship rife with rumors and speculation amongst the officers and men of the two helicopter units and their supporting field maintenance, signal and medical detachments, especially since the San Francisco Chronicle, the day before on 20 November, had already printed the probable destination, a place that most Americans were unaware of, Vietnam! It was only when the ship was underway that sealed orders could be opened and the identity of the Overseas Exercise Area, Vietnam, was officially confirmed.

Three weeks later on 10 December 1961, as the USNS Core arrived in Vietnamese waters at the mouth of the Mekong, Major General Charles J. Timmes, who had succeeded General McGarr as MAAG, Vietnam commander, boarded the ship to officially welcome the cargo helicopter units to Vietnam. Depreservation of the helicopters was initiated while the USNS Core left the Mekong and sailed up the Saigon River to dock at the foot of Tu Do Street, in Saigon. Large crowds had gathered to observe the Piasecki designed, Vertol Aircraft Corporation H-21 Shawnee helicopters, nicknamed 'flying bananas' because of their shape, being readied for flight and then lifting off the ship's flight deck to become the first U.S. Army helicopters to fly in the Republic of Vietnam. The event also attracted the attention of the National Liberation Front. "Within a few days, Radio Hanoi broadcast not only the numbers and types of aircraft that had been brought in, but their serial numbers as well."[1] The destination, for both units, was Tan Son Nhut Airport, which was both a civilian and military airfield serving Saigon, approximately ten miles from downtown, and to the west of the city

Although the 8th Transportation Company (Lt Hel) had arrived in Vietnam short its full complement of aircraft, each helicopter unit, at full strength, was equipped with 20 H-21 Shawnee light cargo helicopters and 2 H-13 Sioux observation helicopters. The H-13's were eventually returned to the United States, "since they could not stay up with the H-21's and had no range for recon or route planning,"[2] and their place was to be taken by 2 to 4 fixed wing TL-19D's (soon to be redesignated TO-1D's) Bird Dogs allocated to each company. After a successful training exercise, to the north of Saigon, with Vietnamese paratroopers, the 8th and 57th combined to conduct an actual combat assault, code named Operation Chopper, on 23 December 1961. This combat assault, which delivered 1,000 ARVN paratroopers into what one pilot called "an almost impossible landing zone: perpendicular rows of pineapple fields, with 3-foot troughs between rows filled with water and mud.",[3] disrupted a Viet Cong headquarters and captured an underground radio transmitter. The landing was not unopposed, the helicopters were brought under small arms fire almost the minute they touched down, and one of the 57th's aircraft was lost in another, less welcome, first for Army aviation in Vietnam. A second joint

Bell H-13E Sioux observation helicopters, s/n's visible 51-14220, 51-13824, 51-14217 and 51-13755, originally assigned to the cargo helicopter companies per TOE, await return shipment to the United States on 18 September 1962. Aircraft are painted in gloss olive drab with the Army's standard high visibility markings. US Army Photograph

operation on 2 January 1962, into a heretofore inaccessible jungle area, amazed and impressed the ARVN command. Vietnamese commanders, and their staff, were beginning to understand the incredible implications of airmobility, such as a freedom of movement they had never even contemplated before, the element of surprise and great tactical flexibility. Demand for helicopter support grew, so to meet rising ARVN expectations, the 8th Transportation Company (Lt Hel) relocated on 10 January 1962 to Qui Nhon to support II Corps.

Even though the 57th and 8th were officially only advising and supporting the ARVN, the fact of the matter was that the enemy made no such distinction, shooting at their aircraft from the start. When the 8th Transportation Company (Lt Hel) sustained the first seriously wounded helicopter crewman, in Kontum Province, and the Army awarded the first Purple Heart to an Army aviator in Vietnam, the men of the helicopter units were becoming aware of a widening gap between reality in the field and official policy in Saigon and back in the United States. While conducting training or actual combat assaults into contested landing zones, helicopter units could only administratively log in their flying hours under the two headings, training or service. Official U.S. government policy was still attempting to reconcile two opposing points of view, President Diem's who had reluctantly accepted U.S. military forces and then only as a means to train the ARVN, and the Kennedy administration's conclusion that to win against communism in Vietnam meant that U.S. forces had to engage in conventional, or unconventional, warfare, a posture that could not be made public. So the publicly stated U.S. position, from the administration, to the Pentagon to MAAG, Vietnam headquarters, was that U.S. forces were in Vietnam to advise and support the ARVN; the easiest course of action when confronted with the dilemma of having to choose between conflicting positions. In the field, helicopter crewmen quickly discovered the disconnect between the 'official' line and reality, for it was painfully obvious that flying into a landing zone where small arms fire could be expected and your own people were getting wounded hardly qualified as just 'training'. While this reality gap may have been subtle in early 1962, it existed and grew throughout the year until early 1963 when it split wide open. The situation was noted at the Pentagon, which, hearing the rumblings from the field, eventually coined the term "combat support" to describe the flying activities of the helicopter units. The crewmen of the helicopter units themselves had no illusions that they were in combat, and had to devise some means to defend themselves. It then took a directive from the Kennedy White House, confirmed on 14 February 1962, to allow the H-21 units to return fire when fired upon.

In the meantime the 57th

Transportation Company (Lt Hel) began an intense, one time priority training program for the ARVN in airmobile operations. Overcoming their initial fear of the noise and vibrations of the helicopters, in the first three months 26,000 ARVN troops were trained in the procedures of a combat assault and the techniques of on-loading and off-loading H-21 helicopters. As a result well-trained ARVN troops could on-load an H-21 in 18 seconds and off-load in 12 seconds, once locally fabricated steps were attached to the H-21 to accommodate the smaller statured Vietnamese soldier.[4]

In December 1961 a third H-21 unit, the 93rd Transportation Company (Lt Hel) left the United States on the USNS Card and sailed to Subic Bay in the Philippines where the unit's aircraft were depreserved and prepared for flight. Then, in mid January 1962, the aircraft were flown aboard the USS Princeton (LPH 5) and the carrier set sail for Vietnam. In order to maintain as low a profile as possible, when the USS Princeton approached the coast of Vietnam, at a point approximately ten miles out to sea, the H-21's were flown off the carrier in adverse weather to their new home station, Da Nang Air Base. The 93rd Transportation Company (Lt Hel) was declared operational on 1 February 1962 and began to provide airmobile and logistical support to the ARVN and U.S. Special Forces in I Corps.[5]

At that point in time, February 1962, the Republic of South Vietnam was divided into three military corps regions, I Corps in the north, II Corps comprising the central region of the country, and a large III Corps which included the Mekong River Delta region and extended far north of Saigon. The 93rd Transportation Company (Lt Hel) supported I Corps from their base at Da Nang, II Corps was supported by the 8th Transportation Helicopter Company (Lt Hel) in Qui Nhon and III Corps support was provided by the 57th Transportation Company (Lt Hel) based at Tan Son Nhut, Saigon. General Taylor's recommendations had been enacted, three helicopter companies were now in place, supporting the three corps areas.

A joint planning session for the first heliborne assault in Vietnam in December 1961. Clockwise from top left: Major Robert Dillard, Commanding Officer, and Captain Emmett Knight, Operations Officer, both from the 57th Transportation Company (Lt Hel), Lieutenant Colonel Allen Goodson, Senior US Advisor to the RVN Airborne Brigade, Major An, S-3 of the RVN Airborne Brigade, ARVN unknown, Lieutenant Colonel John B. Stockton, MAAG Advisor, and two more ARVN unknowns. Captain Knight, together with CWO Bennie Potts, flew the lead ship in Army aviation's first combat assault of the Vietnam War. Lieutenant Colonel Stockton did not endear himself to the first Army aviators, or their wives, when he was quoted in Time magazine, after this first operation, as saying "It was just like World War II out there."

Photograph courtesy of Col. Emmett Knight, USA Ret.

One of the first tasks given to the 93rd Transportation Company (Lt Hel) was to reestablish contact and communications with isolated ARVN outposts within I Corps. So successfully did the 93rd perform this mission that there was a marked improvement in ARVN morale. Dependable aerial resupply allowed the ARVN to establish new outposts and encouraged aggressive patrolling in areas previously conceded to the enemy, such as the A Shau Valley. While the A Shau Valley was seldom controlled by South Vietnamese government forces, and was often in enemy hands for extended periods, its importance as an infiltration route justified every effort by the South Vietnamese to patrol it and limit its use to the enemy .

In January 1962, the 45th Transportation Battalion (Transportation Aircraft) had been ordered to Vietnam from Fort Sill, Oklahoma to assume command of the three H-21 helicopter units. The 45th Transportation Battalion became the second Army aviation unit to be based at Tan Son Nhut Air Base. As the 45th Transportation Battalion was settling in, they had been formed from scratch at Fort Sill, and it took, in reality, a few months before they actually had much to do with the units under their command, the training of the ARVN in airmobile operations by the H-21 units, as well as

A 57th Transportation Company (Lt Hel) H-21C, s/n 56-2019, in gloss olive drab and the Army's standard high visibility markings with white flight platoon colors, serves as a training aid for ARVN infantrymen in the proper procedures to on-load and off-load helicopters. CWO Richard Parson was the US instructor in March 1962.

US Army Photograph

the conduct of actual airmobile assaults, continued in spite of maintenance difficulties and an almost non-existent supply system. The 9th Logistic Command in Okinawa (that island chain being known as the Ryukyu Islands prior to being returned to Japanese sovereignty in the mid 1970's) had been directed to send a logistical support team to Vietnam in late 1961. Their mission was to establish, and have up and running, a supply system service between the Army aviation units in Vietnam and U.S. Army, Ryukyu Islands (USARYIS). From this small beginning developed the U.S. Army Support Group (Provisional) which was the forerunner of the U.S. Army Support Command, Vietnam and later, the U.S. Army, Vietnam (USARV). However, as the fledgling USARYIS Support Group (Provisional) was being formed "resupply of aviation parts and supplies was extremely critical during the first few months while a supply system was being established. Initially, supply support was the responsibility of activities located in Okinawa, but the short notice and the distance removed combined to produce a far-from-adequate supply system. Eventually, an Aviation Supply Point was established at Tan Son Nhut and became minimally effective during the early months;"[6] That the H-21 units accomplished their mission was due to both the considerable professional ability of Army pilots and the highly skilled, motivated, and long-suffering maintenance personnel who worked day and night to keep their ships in the air. While the pilots might 'borrow' an aircraft to fly, the crew chief 'owned' it, for the Army had assigned each aircraft to a crew chief, thus instilling strong pride of ownership. The crew chief was responsible for 'his' aircraft, and worked hard servicing, cleaning, inspecting and checking that write-ups were cleared from logbooks. The crew chief would often 'name' the aircraft and have his personal marking applied by a unit member or a local Vietnamese artist in a place where all passengers would be sure to see it. The 57th Transportation Company's (Lt Hel) unit commander, in 1963, even established a unit tradition to provide a free drink to the crew chief and gunner of any aircraft that was unlucky enough to be hit by enemy fire.

The U.S. Military Assistance Command, Vietnam (MACV) was officially established on 8 February 1962 under the command of General Harkins as had been announced by President Kennedy in November 1961. The former MAAG Flight Detachment, which was the first Army aviation unit in Vietnam when established in October 1954 with four aircraft, comprised of L-20A Beavers and L-23 Seminoles, was redesignated the MACV-MAAG Flight Detachment. The MACV-MAAG Flight Detachment flew their immaculate red and white VIP aircraft in support of the two command headquarters.

In March 1962 the "provisional" designation was removed from the U.S. Army Support Group, Vietnam which was still subordinated to USARYIS for administrative and logistical support, but was commanded by the deputy Army commander in MACV. While all Army units in Vietnam were

ARVN infantrymen get hands-on training and practice on-loading H-21C, s/n 56-2049, of the 57th Transportation Company (Lt Hel) in early 1962. The locally fabricated step facilitated this process considerably.

US Army Photograph

assigned to the Army Support Group, Vietnam operational control of all aviation companies remained with General Timmes, MAAG, Vietnam. "Control of Army aviation assets at this time illustrates the multiple lines of responsibility in Vietnam. Since General Timmes had operational control of Army aviation units, the senior adviser assigned to a Vietnamese Army corps could directly request U.S. Army aviation support. For example, the Vietnamese corps commander could initiate and plan a helicopter operation. The adviser assigned to the corps would formally transmit a request to the commanding officer of a U.S. Army helicopter company for execution. Actual planning for such an operation thus involved the Vietnamese corps commander and his staff, the Military Assistance Advisory Group's representative, and the commander of the helicopter company. ....Army aviation unit commanders, therefor, had to deal with and satisfy, on a daily basis, the Vietnamese Army, the Military Assistance Advisory Group, the Military Assistance Command, and the U.S. Army Support Group."[7] This command relationship was so involved that it is a tribute to Army aviation unit commanders, not that it worked so well, but that it worked at all!

In February the 18th Aviation Company (U-1A) with its fixed wing Otters was delivered by the USNS Core, and became operational in March after its aircraft had been reassembled at Tan Son Nhut by mechanics of Air Vietnam, the Vietnamese national airline.[8] The 18th Aviation Company (U-1A) soon relocated its headquarters to Nha Trang, while the unit's flight platoons were stationed where they could best support U.S. Special Forces units and all the corps areas. 'Low, Slow and Reliable' soon became more than a company motto, as the Otters quickly proved their value in establishing a utility air-net capability uniting the widely spread aviation units with their support elements. Also in February the 339th Transportation Company (Direct Support) had been deployed to Nha Trang and was tasked with the maintenance support of the Otters of the 18th Aviation Company as well as the almost impossible mission to provide back-up support throughout the country to the Cargo Helicopter Field Maintenance (CHFM) detachments assigned to each helicopter unit.

The deployment of aviation units to Vietnam was beginning to accelerate. In March 1962 the 3rd Radio Research Unit (RRU) 'borrowed' three L-20A (U-6A) aircraft, possibly from the MACV-MAAG Flight Detachment, and fitted them with airborne radio direction finding (ARDF) gear. These three ARDF equipped aircraft were then stationed at Tan Son Nhut. This radio intercept intelligence gathering unit was assigned to the Military Assistance Command, Vietnam but was staffed by the low profile Army Security Agency and reported to the National Security Agency at Fort Meade, Maryland. The Aviation Section of the 3rd Radio Research Unit was officially organized in June 1963 and flew specially modified L-20A Beavers and RL-23D Seminoles configured for electronic intelligence gathering. While never widely publicized, the Army Security Agency crew members, they and their aircraft often referred to affectionately as 'spooks', of the 3rd RRU Aviation Section were responsible for eavesdropping on NLF radio traffic and for identifying the location of clandestine Viet Cong radio transmitters.

On 6 March 1962 a fourth Army helicopter unit, the 33rd Transportation Company (Lt Hel), at Fort Ord, California, was alerted to prepare for departure to Vietnam on 18 April. General Harkins had, in the meantime, requested a Marine Corps Helicopter Squadron instead, setting into motion a plan, to have Marine Corps pilots augment the Army's, that had been suggested by General Timmes of MAAG, Vietnam. This complemented the secret contingency plans then in place which assigned the Marine Corps a sector of the country if the North Vietnamese should invade the South. The 33rd Transportation Company stayed home, for the time being, and a Marine helicopter unit in Okinawa was alerted for movement.

Occurring in April of 1962, this U.S. Marine H-34 helicopter squadron was deployed to Soc Trang in the Mekong River delta, known simply as 'the Delta'. The Army 93rd Transportation Company had just set up in I Corps, so the Marines were temporarily sent to the Delta, until a switch could be made. Helicopter Marine Medium 362 (HMM-362), nicknamed "Archie's Angels" after their commander, is worth mentioning, within the context of an Army aviation history, for the unit's role in developing the famous 'Eagle Flight.' The 'Eagle Flight' concept had its origins in an operation on 19 May 1962 when an attempt was made to capture some Viet Cong leaders in a village near Soc Trang. While ARVN troops were landed at each corner of the village, most of the Viet Cong were able to evade capture and made their way towards a river some distance away. At this point, four helicopters, which had been held in reserve and out of sight, were brought in and landed between the fleeing Viet Cong and the river. The success of this operation soon led to the reserve force being used by itself, cruising around looking for the enemy. The tactic was the inspiration of HMM-362's commander, Lt. Col. Archie J. Clapp, and was later to be described as "a tactical concept involving the employment of a small, self-contained, highly-trained heliborne force......... The inherent flexibility of the 'Eagle Flight' as a force ready for immediate commitment either alone or in conjunction with other forces is its most significant factor."[9] The enemy, who quickly analyzed any new tactics, referred to the 'Eagle Flight' tactic as 'Eagle catches chickens'; presumably the local Viet Cong being the chickens![10] So seriously did the enemy view this development that regular English language broadcasts from Radio Hanoi, North Vietnam, soon labeled the unit " the Marine butchers of the 362nd" and theorized, for good measure, that they ate human liver for breakfast! [11]

While the 'Eagle Flight' concept may have been conceived by HMM-362, it had its antecedents within the H-21 units. The contributions of these pioneer Army aviators within the first year and a half of U.S. military involvement in Vietnam was never well publicized at the time and has received but brief mention in subsequent histories. The commanders and operations officers of early Army aviation units in Vietnam were professional Army pilots, but, more importantly were, in the truest sense of the word, warriors. From day one they had used all their prior knowledge, military experience, training, ingenuity and creativity to execute their assigned mission successfully. Tactics were assessed and reassessed as conditions and enemy response changed. This evolution in tactics can clearly be followed in the flight patterns tested and refined as time went on. At first combat assaults were flown at contour level from troop pick-up to the landing zone, in order to achieve surprise and create maximum shock effect. This tactic was soon modified for any subsequent return flights, with additional troops, no longer had the element of surprise, and exposed the aircraft to enemy fire along the route. A safe minimum altitude of 2,000 feet was established for flights to pick-up troops, and enroute to the landing zone, until, at approach, helicopters dropped to contour level and roared in low to the ground to achieve surprise. By 1963 the minimum safe altitude had been raised to 3,000 feet as the Viet Cong's arsenal of automatic weapons, many captured from the ARVN, or scavenged from downed aircraft, had grown. Successful navigation at contour level was difficult, and the need for a higher flying, directional controller became apparent. Initially fixed wing aircraft, such as the L-20A Beaver, of the MAAG Fight Detachment were borrowed to fill that function until the units acquired their own TO-1D Bird Dogs. Routes in and out of a combat assault, and the landing zone itself, were selected carefully to avoid their use more than once thus denying the enemy a chance to execute an ambush. These, often unheralded, Army aviators in 1962 and 1963 were writing the first chapter, line and verse, under combat conditions, of the concept that became known as 'airmobility'. As they finished their tours in Vietnam and went on to other positions, stations and commands the lessons they had learned, the tactics they developed and their experience in a new type of warfare built a solid foundation upon which Army aviation could develop in the years ahead in Vietnam.

Also in April 1962 a small medical detachment, the 57th Medical Detachment (Helicopter Ambulance) was stationed in Nha Trang with five UH-1A Hueys. The rational for this posting was that the 8th Field Hospital had just been established in Nha Trang, although this command was totally unaware that the 57th Medical Detachment (Hel Amb) even existed. In this first, tentative step, a remarkable and heroic tradition within Army aviation was born in Vietnam, which,

H-34's of Helicopter Marine Medium 362 (HMM-362) prepare to touch down in a rice paddy in mid 1962. These aircraft were painted in Field Green (FS. 34097) and carried the National Star Insignia, white MARINE, unit designation and BuNo lettering and yellow danger tail marking with red arrow and lettering. HMM-362 painted their aircraft's engine covers yellow as a unit marking. This U.S. Marine Corps unit came in for particularly vehement attention from Radio Hanoi. US Army Photograph

to this day, has become known simply as "DUST OFF". However the 57th Medical Detachment (Helicopter Ambulance), through no fault of its own, got off to a slow start as Nha Trang was too distant from regions where sustained combat action was occurring such as in the Delta or in I Corps. The unit was almost put out of business in late 1962, when Huey starter generators were so desperately needed that the 57th's five Hueys were ordered cannibalized, and then only one starter generator was returned so that the 57th Medical Detachment (Hel Amb) began 1963 with a single flyable helicopter ambulance! At one point it was being seriously debated whether or not the unit should be disbanded and temporary 'red cross plaques' be carried by all aircraft, turning any aircraft into an instant helicopter ambulance when needed. It is a testament to the spirit and persistence of the 57th Medical Detachment (Hel Amb), 'the Originals', that they survived, and through determination got themselves a change of stations from Nha Trang to Tan Son Nhut to support operations in the Delta, where they more than demonstrated their effectiveness, they proved beyond a doubt their right to exist.

Meanwhile, the efforts of the 93rd Transportation Company (LT Hel) in I Corps had not gone unoticed as General Hightower, on a country-wide tour, selected the 93rd Transportation Company, and its attached units, as the most outstanding company that he had visited in Vietnam. Recognition also came to the 45th Transportation Battalion (Trans Acft) from the Hughes Tool Company, Aircraft Division which awarded the 45th the 'Outstanding Aviation Unit Award' in 1962 inscribed, in part " ... for making outstanding contributions to the employment of Army Aviation over and above normal assigned mission thereby helping chart the path of tactical mobility while assisting a valiant ally to survive as a free nation." [12]

In July 1962 the 1st Platoon of the 1st Aviation Company (CV-2) had arrived in Vietnam with its 8 De Havilland Aircraft CV-2 Caribou. The entire 1st Aviation Company had deployed in May 1962 to Korat, Thailand from Fort Benning, Georgia in support of Joint Task Force 116. The 1st Platoon was detached, as mentioned, and its aircraft were located to best serve each corps area, 2 aircraft in Da Nang, 2 aircraft in Nha Trang and 4 aircraft located at Tan Son Nhut Air Base. It should be noted that a single YAC-1 Caribou, s/n 57-3080, being tested by the Advanced Research Project Agency (ARPA), had met, or exceeded, all expectations in Vietnam as early as August 1961, remaining in-country through 1964.

The UH-1A Flight Section also arrived in July and was attached to the 45th Transportation Battalion (Trans. Acft). The unit was tasked with supporting MAAG advisors in I, II and III Corps.

The 93rd Transportation Company (Lt Hel) ushered August in by participating in the first large scale heliborne operation ever conducted in I Corps when two companies of ARVN raided the Viet Cong's 5th Region Headquarters. Following up this successful operation was the joint airmobile combat assault conducted on 30 August, 1962 which paired the 93rd with the Vietnamese Air Force's (VNAF) 1st Helicopter Squadron (H-34). Code named Operation 'Lam Son II' the intent was to insert troops on top of a regular Viet Cong battalion immediately after tactical air strikes had been delivered. While the air strikes were delivered as scheduled, poor weather delayed the helicopter airlift. The enemy, it was later learned, had previous intelligence information on the operation, and used the delay to recover from the air strikes and to prepare themselves for the combat assault. Although the 93rd completed two airlifts into the landing zone, nine H-21's were hit and two were downed by heavy enemy fire. Despite these losses, Lam Son II was considered successful in that many of the enemy had been killed, valuable intelligence documents and prisoners were taken and supplies were destroyed. Although it was not apparent at the time, this compromised opera-

As the 57th Transportation Company (Lt Hel) H-21's touch down, ARVN troops off-load to conduct a combat assault on 20 December 1962. H-21C, s/n 55-4165, has had its high visibility markings painted over and carries the name 'The Hedge Hopper.' US Army Photograph

Initially door gunners came from within the ranks of the unit, and its attachments. Aircrew in a H-21, consisting of the pilot, co-pilot, crew chief and gunner, formed a team; the gunners being well respected within the unit for their dedication and in recognition of the hazardous nature of their job. Most of the H-21 units designed 'gunner's badges' that qualified gunners could proudly wear on their hats. Later on the need for gunners was met by sending twenty man platoons, with one officer, to Vietnam from the 25th Infantry Division in Hawaii. These 'Gunner Platoons', as they were called, were trained in Hawaii, were experts with their weapons and manned and cared for the machine guns on the H-21's. These platoons were rotated through the H-21 units, and later the UH-1 Air Mobile units, on a ninety day temporary tour of duty.[18]

Additional fire suppression capability, other than the door guns, was needed and at least two of the H-21 units, the 81st Transportation Company (Lt Hel) and the 57th Transportation Company (Lt Hel) experimented with the XM-1 sub-system, originally designed for the H-13 and H-23, on one or two of their aircraft. Affixing machine guns to the nose of the H-21 was an unsatisfactory arrangement, and the experiment did not catch on. However the introduction of the armed UH-1A's, and later in November 1962 the UH-1B's, of the Utility Tactical Transport Helicopter Company (commonly referred to as the UTT) provided the best available solution to the problem. It is significant to note that the number of VNAF helicopter escort sorties, which had totaled almost 700 between August and December of 1962, dropped to half of that, 350 helicopter escort sorties, between January and May 1963. The introduction of armed helicopters into combat was, to say the least, disturbing to U.S. Air Force (USAF) staff, who foresaw a usurpation of their traditional role. Enthusiastic Army aviators and the press did not help the situation much as contemporary accounts labeled the armed Hueys of the UTT as 'fighters', 'flying tigers' and the 'Cadillac's of the air'. The USAF first attempted to define the role of the armed helicopter in December 1962, and continued to monitor the Army's use of armed helicopters throughout 1963, objecting strenuously when it appeared that the Huey's were used 'offensively.' The Army successfully countered these allegations, which in the words of one participant were "so ridiculous it was humorous. As if there were not enough enemy targets for all available weapons, ground or air."[19] In the opinion of one of the early members of the UTT "for considerable periods of time the UH-1B's of the UTT represented the most effective fire support available in the remote area of III, IV and I Corps. ...they, more than any other means of fire support, allowed the early days of airmobile assault in Vietnam to be possible."[20]

Besides members of the unit themselves, much of the credit for the success enjoyed by the UTT company, and for the development of armed helicopter tactics as a whole, was due to the Army Concept Team in Vietnam (ACTIV) headed by Brigadier General Edward L. Rowny. ACTIV's first aviation representatives arrived in Saigon in September 1962, and, although not solely concerned with Army aviation or aviation tac-

Two Hueys of the UTT can just be made out as they overfly H-21's about to touch down in a LZ. Army aviators of the cargo helicopter companies came to prefer the UTT escorts over the VNAF T-28's, to the consternation of the USAF.

Photograph by Captain Roger Baker via Bob Chenoweth

H-21's in a LZ photographed from an escorting Huey of the UTT. At least one of the H-21's is from the 93rd Transportation Company (Lt Hel) as it carries a red conspicuity stripe on the top of the fuselage.

Photograph by Captain Roger Baker via Bob Chenoweth

tics, ACTIV was instrumental in the introduction of new aircraft types and weapon systems. The UTT company's armed helicopter's effectiveness was evaluated by ACTIV from October 1962 through March 1963. Subsequently ACTIV evaluated the effectiveness of camouflage on the helicopters of the UTT company, the UH-1A's having all their high visibility markings removed almost immediately upon arrival in Vietnam, while the UH-1B's were operated in the standard Army high visibility paint scheme. It was in mid 1963, and into 1964, that some of the UTT's UH-1B's were tested using two and three-color camouflage paint schemes. It should be noted that the H-21 units had already begun to tone down the high visibility markings of their aircraft by October 1962, painting over the large, white ARMY marking and the 'stars and bars' national insignia on the fuselage, since it was felt that these markings provided too good an aiming point for the enemy. Additionally, some units began to camouflage their helicopters, starting with the 8th Transportation Company (Lt Hel) in Qui Nhon in December 1962 and spreading to the units in the Saigon area, so that, by mid 1963, many H-21's were sporting a unique three-color camouflage scheme, consisting of blue-green and tan over the basecoat of olive drab.

The U.S Army Support Group, Vietnam, commanded by Brigadier General Joseph W. Stilwell Jr., was tasked with the administrative, combat and logistical support of all Army activities, except advisory attachments. Brigadier General Joseph W. Stilwell Jr.'s role in supporting and encouraging Army aviation's early experiments and operational testing of weapons and tactics in Vietnam has not received the full mention and recognition that it deserves. General Stilwell, with all the aviation units under his command, decided that he too should be qualified as an Army aviator, and proceeded to learn to fly 'bootleg', taking flying lessons from the U.S. Army Support Group Aviation Officer. After he gained his wings Army aviation units anywhere could expect a visit from the 'Flying General' in his L-19 Bird Dog. That is not to say that the sight of the General landing and taxiing up to the Operations building was an eagerly awaited experience. General Stilwell had a reputation as a " difficult man to work for, he was profane, arrogant and tried to live up to his namesake and father, Vinegar Joe Stilwell."[21] He was a tough taskmaster and woe betide the luckless unit commander who failed to live up to the General's expectations. However, with 'Cider' Joe's enthusiastic support and high standards, assisted by his Inspector General (IG) Lt. Col. Kenneth D. Mertel, acting as the General's eyes and ears in the field, Army aviation units, building on their experiences the year before, had established such a high level of excellence and esprit by 1963 that it would serve as a

Right: **General Stilwell, who often flew as a door gunner in the UTT's Hueys, takes a breather in Soc Trang in May 1963. To his right is 1st Lieutenant James E. Stone of the UTT who was awarded the DFC for heroism at Ap Bac on 2 January 1963.**

Photograph courtesy of Col. Sam Conley, Jr., USA Ret.

Below: **An informal photograph of Brigadier General Joseph W. Stilwell, Jr., Commander of the US Army Support Group, Vietnam and his Inspector General, Lieutenant Colonel Kenneth D. Mertel in mid 1963.**

Photograph courtesy of Col. Kenneth D. Mertel, USA Ret.

Below: **Lieutenant Colonel Lester C. Robertson, US Army Support Group Aviation Officer, and General Stilwell's Instructor Pilot, at Vung Tau in March 1963.**

Photograph courtesy of Col. Sam Conley, Jr., USA Ret.

benchmark against which all subsequent Army aviation units in Vietnam would have to strive to meet, let alone surpass.

The Army had the opportunity to evaluate another armed aircraft when six Grumman JOV-1C Mohawks of the 23rd Special Warfare Aviation Detachment (23rd SWAD) arrived at Nha Trang in October 1962. The armed Mohawks of the 23rd SWAD flew in support of ARVN forces in II Corps. At first the Army's 'rules of engagement' limited the Mohawk aircrews to returning fire only if fired upon, and restricted this return fire to .50 caliber weapons. Later, after relaxation of the rules of engagement in early 1963, the Mohawks were armed with rocket pods and they had an effect on the enemy quite beyond their numbers, causing commanders at all levels to press for close air support from the Mohawks. The detachment was eventually relocated to Vung Tau but positioned 2 aircraft sections, at various times, at Qui Nhon, Soc Trang and Pleiku.[22] One of the 23rd SWAD's Mohawks mysteriously disappeared on a routine mission out of Pleiku in February 1963, and was not to be discovered until December of that year when spotted by a Bird Dog of the 73rd Aviation Company. The cause of the crash was never determined. The 23rd SWAD is known to have lost at least two more Mohawks in March 1963; in the first incident, the crew were able to eject, while, in the crash of an aircraft flying out of Soc Trang, the crew were killed.

In December 1962, after the conclusion of the exercise Joint Task Force 116 in Thailand, the 1st Aviation Company (CV-2) rejoined its 1st Platoon, which had been flying in Vietnam since the previous July. Headquartered at Vung Tau, the Caribou's of the 1st Aviation Company were quickly put to work. The 1st Aviation Company, nicknamed 'Pathfinder', successfully originated an air net, tying together airstrips which ranged from hard surface to sod and gravel, most of which were denied, due to their limitations, to any USAF cargo aircraft. Soon units everywhere in Vietnam, from MAAG Detachments and U.S. Special Forces outposts to Army aviation outfits throughout the country, had begun to rely on, and take for granted, the regularly scheduled "milk runs" of the Caribou's. The 1st Aviation Company (CV-2) remained in Vietnam until 7 December 1963 when the unit was returned to the United States as part of a token force draw down announced by the Kennedy administration, and, in another first for Army aviation, was the first Army aviation unit to fly its organic aircraft home from overseas deployment.

Captain Walter K. Rayal, of the UH-1A Flight Section, 45th Transportation Battalion, gives scale to the sharpened 'anti-helicopter' poles placed by the Viet Cong in a potential helicopter landing zone near Tuy Hoa, II Corps. The date is 17 October 1962.
US Army Photograph

For most of 1962 the enemy had been caught off-balance by the new mobility of the ARVN. Enemy tactics had generally precluded standing and fighting, but clearly both active and passive resistance would be necessary to counter airmobile tactics and restore morale to increasingly discouraged forces. The active measures that the enemy took included training in the necessity of 'leading' a flying target. In the case of the H-21, which the Viet Cong (VC) deprecatingly called 'Angle Worm' because it was large and looked like a worm, they recommended leading by a length or two thirds of a length in horizontal flight; the smaller UH-1, nicknamed 'Dipper' by the VC because it looked like a ladle, the enemy recommended leading by a length, and firing into its engine when it was at hover or landing. Passive measures employed were primitive, but often effective, including placing sharpened poles in potential landing zones and a crude bow and arrow system that was triggered by the rotor wash of a landing helicopter.[23]

The first significant challenge to airmobile tactics was made by the enemy at a battle near the hamlet of Ap Bac. A turning point in the war for Army aviation, this engagement began routinely, a seemingly typical "clean and sweep" lift for the 93rd Transportation Company (Lt Hel) and the escorting UH-1B's of the UTT Company. Late in December 1962, an Army Security Agency crew in an aircraft of the 3rd RRU Aviation Section, monitoring enemy radio traffic in the Delta, pinpointed an enemy transmitter near the hamlet of Ap Bac. Since a radio transmitter was part of the Viet Cong's command and control infrastructure, locating a transmitter often meant that an enemy headquarters was positioned close by.

On 28 December the ARVN 7th Infantry Division was alerted and ordered to seize the transmitter. The ARVN 7th Infantry Division commander Colonel Bui Dinh Dam, and his U.S. Army adviser Lt. Col. John Paul Vann, immediately began to plan an airmobile assault, assisted by ground forces, to eliminate this enemy unit, which the best intelligence estimates considered to have a strength of no more than 120 men. The 7th Infantry Division was to provide three battalions for the mission, augmented by ARVN Rangers and two battalions from the paramilitary forces of the Civil Guard. In addition an infantry company mounted in M-113 armored personnel carriers, artillery, two U.S. helicopter companies and the VNAF were notified to prepare to support the operation.

The operation was compromised, through enemy infiltration of the government and ARVN command structure, and the local NLF party chief and his battalion commander were alerted that an attack was imminent. Ap Bac was a 'liberated' village, occupied at that time by the Viet Cong 261st Main Force Battalion, with a company from the 514th Regional Battalion, reputed to be among the best enemy units in the Delta. These units could take advantage of previously prepared positions and bunkers along the dikes and tree lines at the edges of the hamlet. These bunkers had been constructed in most Viet Cong controlled localities, and were, more often than not, designed and built with a keen appreciation of airmobile tactics. At Ap Bac the bunkers were located along raised dikes and in the heavy undergrowth of tree lines with good overhead cover, were well camouflaged and had excellent fields of fire across open rice paddies. The two enemy units, the 261st Main Force Battalion and the detached company of the 514th Regional Battalion, defended Ap Bac with approximately 320 troops together with some local and village forces. The enemy, waiting at Ap Bac, was composed of two units that were extremely well disciplined and combat hardened.

Postponing the operation by one day, originally the assault was scheduled for 1 January 1963, Col. Dam considerately allowed the helicopter crews a chance to celebrate the New Year. Lt. Col. Vann had requested that more helicopters be made available than the ten committed by the 93rd Transportation Company (Lt Hel). His attempts to get additional lift capability were in vain since other units, such as the 57th and 33rd, were assigned to support a higher priority mission which was taking place northwest of Saigon in War Zone C. This Operation, code named Burning Arrow, was an attempt to neutralize the enemy's highest command, the elusive Central Office for South Vietnam (COSVN), but it failed, proving to be no more successful than other efforts to eliminate COSVN. Besides conflicting priorities, aircraft availability was dependent on logistical support, and despite the long hours and the best efforts of truly dedicated maintenance personnel, supply problems had begun to seriously erode unit readiness. 'Boneyards' where crashed or battle damaged aircraft were stripped of usable parts to supplement the supply system, could not make up for the critical shortages in engines, rotor blades and avionics equipment. It was not uncommon to task three cargo helicopter companies to guarantee enough aircraft to support even routine missions.

In the early hours of 2 January 1963 the Civil Guard forces moved into place, approaching Ap Bac from the south. At dawn the artillery commenced preparatory fires. The enemy was so well dug in, and camouflaged, that not only were they not visible, either from the air or the ground, but it required a direct hit from a shell, or rocket, to destroy their positions. As the artillery preparatory fires lifted, the first attack was launched by the Civil Guard across the rice paddies against the southern end of Ap Bac. The attackers were slowed by knee deep mud, and the entrenched enemy halted this and subsequent attempts by the Civil Guard to advance. The Civil Guard commander, Major Lam Quang Tho, then requested that the 7th Infantry Division

While primitive and crude, the enemy's sharpened 'anti-helicopter' poles, faintly visible scattered throughout the fields, could only be ignored at great risk as this photograph illustrates on 8 December 1962. The H-21, at left, crashed, and later had to be destroyed, after colliding with enemy poles. The crews of the other helicopters have landed to provide assistance in removing all useable equipment so that it would not fall into enemy hands.

US Army Photograph

reserve troops, waiting at their airstrip in Tan Hiep, be airflifted in and landed behind the southern tree line from which he was taking so much fire. This envelopment was supposed to cause the Viet Cong to abandon their positions and melt away, as they often had in the past, when confronted with superior forces and airmobile tactics.

Lt. Col. Vann was flying over Ap Bac, as the backseater in a L-19 Bird Dog, to observe the battlefield and direct the operation. In spite of heavy ground fog, the ten H-21's of the 93rd Transportation Company (Lt Hel) on-loaded fresh ARVN troops at Tan Hiep and airlifted them to Ap Bac. The first lift touched down at approximately 0730 hours to the northwest of the stalled Civil Guard attack and off-loaded troops without opposition. Again the 93rd returned to Tan Hiep, but this time the ground fog, which had thickened thereby raising the risk of midair collisions, delayed the second lift's take off for approximately two hours, until the sun had dissipated the fog. The second and third lifts were then flown in and also off-loaded uneventfully. When the fourth lift, guided by one of the 93rd's Bird Dogs flown by CWO Frank Baldwin, finally returned to Ap Bac, Lt. Col. Vann ordered them in closer than before, to a landing zone approximately 300 yards from the hamlet's western most tree line. Either not hearing the order or determined to choose their own landing site, the lead helicopter 'Chalk 1' (so called from the numbers, marked in chalk beside the main entrance door, used to assign the aircraft's position within the flight) approached a landing zone approximately 200 yards from the tree line. It was the opportunity for which the enemy had been patiently waiting; when the ten H-21's of the fourth lift touched down, in ones and twos, the tree line erupted in accurate and sustained fire. The first four aircraft, even though under heavy fire and sustaining multiple hits, were able to off-load their troops and hurriedly take-off. 'Chalk 5' (aircraft 'Y' s/n #51-15908) was not so fortunate for, as the ARVN troops off-loaded, massed enemy fire succeeded in disabling the helicopter's transmission. As the aircraft's rotors slowed to a standstill, the aircraft commander, Lieutenant Lew Stone, who ironically enough had a helicopter shot out from under him during Operation Lam Son II in I Corps, followed unit procedure and called the next chalk to come get him and his crew. 'Chalk 6' (aircraft 'D' s/n #51-15903) flown by Chief Warrant Officers Carl Flippin and Carleton J. 'Joe' Nysewander unhesitatingly returned and flared into the landing zone, positioning their aircraft between the enemy and the stricken ship in an effort to protect Lt. Stone's crew as they scrambled aboard. This new target attracted so much enemy fire that it too was quickly disabled, losing its aft controls and radio. After crew chief, SP4 Donald Braman, was mortally wounded, CWO Nysewander ordered "Abandon ship!' before he too was seriously wounded.[24] Now two helicopters were down in the landing zone; two others of the fourth lift had sustained so much damage that they had to be set down some distance from the landing zone, the crews being quickly recovered.

The ARVN 7th Infantry Division airstrip at Tan Hiep 20 miles south of Saigon.

US Army Photograph

The armed UH-1B's of the UTT had reacted as soon as the enemy had taken the 93rd's fourth lift under fire and immediately formed a circular 'daisy chain' formation, each covered by the ship behind it, and began to strafe the tree line with rocket and machine gun fire. This type of strafing attack was normally devastating enough to suppress all ground fire, however, at Ap Bac, the enemy stood their ground, undiminished fire being directed back at the Hueys throughout their strafing runs, all aircraft taking hits. The UTT pilots were both surprised and shaken by this aggressive reaction of the enemy and a grudging respect developed for their foe. A fifth H-21, attempting to land, was driven off by enemy fire, but was so badly damaged that the crew were barely able to fly a mile before being forced down. The crew was rescued. Lt. Col. Vann was orbiting overhead, furious that he had five aircraft down and a stalled attack. His mood did not improve when he heard the UTT platoon lead Huey, piloted by Chief Warrant Officer Richard Watlack and Lieutenant James E. Stone (no relation to the 93rd's Lt. Lew Stone stranded on the landing zone) radio that they were going down to extract the 93rd's helicopter crews. As the remainder of the UTT Hueys again strafed the tree line, attempting to provide some cover for this rescue attempt, the platoon leader's aircraft approached from the west. Picking a spot some fifty yards behind the downed ships, the pilot slowed in preparation for landing, and was taken under fire, one round striking the main rotor blade which resulted in the aircraft flipping over onto its right side and crashing, killing the crew chief Sergeant William L. Deal.[25] Six helicopters were down, and American helicopter crews, with dead and wounded, were still stranded on the battlefield.

To rescue the Americans and regain some momentum, Lt. Col. Vann demanded that the ARVN move up the M-113 armored personnel carriers that were supporting the operation. Unfortunately the politics of survival in South Vietnam became apparent when the M-113 company commander Captain Ly Tong Ba would not advance. It was not a question of bravery, for Captain Ba had previously shown himself to have been both a capable and brave leader. Captain Ba was a Buddhist, already under suspicion by President Diem, who was holding up his promotion to Major. Captain Ba knew that a debacle, or heavy casualties would be unacceptable to the Saigon government and would ruin his career. In light of that, Captain Ba managed to stall his advance, infuriating Lt. Col. Vann, who, at one point, ordered the unit's U.S. Army adviser to "..shoot that rotten, cowardly son of a bitch right now and move out."[26] The adviser, Captain Robert Mays, found a way to ease the situation and get the M-113's moving, if only circuitously, on their way to Ap Bac.

Meanwhile Lt. Col. Vann had returned to the Tan Hiep airstrip to refuel the L-19 and plan another attempted rescue of the helicopter crews at Ap Bac. In discussions with pilots of the 93rd it was decided that Lt. Col. Vann would fly back and forth in front of the tree line to 'troll' for enemy fire. If he drew no fire, then the remaining UTT Hueys, only three were still airworthy by this time, would strafe the tree line once more as a two ship H-21 element would approach, from the west; one orbiting at altitude in reserve, while the second flew in low, attempting to use the downed helicopters to shield its approach. As the plan unfolded, and the rescue H-21 touched down, it too was subjected to intense enemy fire, in spite of the UTT's best efforts to suppress that fire, and the crew were forced to immediately lift-off, staggering to a spot about three quarters of a mile away where it was set down, on top of the approaching column of M-113's.

A record seven helicopters

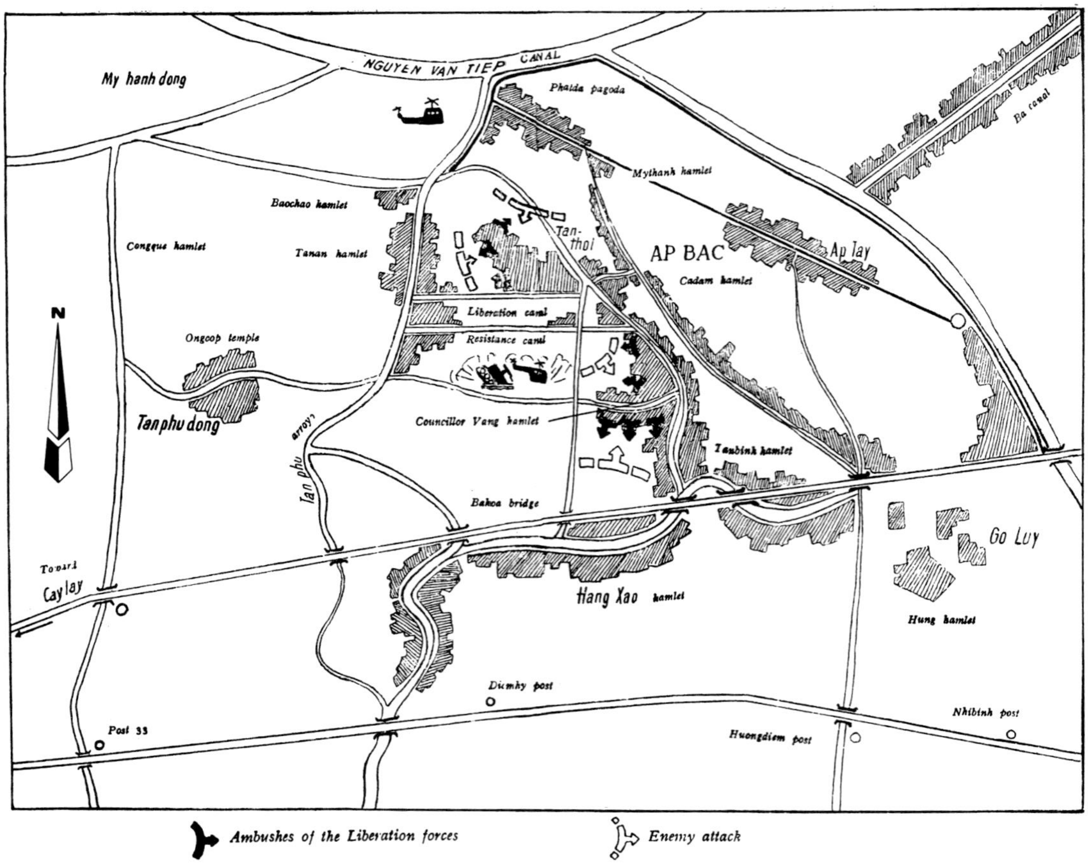

Figure 4.
Diagram of the battle of Ap Bac, reportedly by the NLF, from 'Ap Bac Major Battles in South Vietnam 1963-1964' Foreign Languages Publishing House, Hanoi 1965. The fourth lift's LZ, to the west of Ap Bac, is where 2 H-21's and the UH-1B were downed and the ARVN M-113's stopped.

were now downed; American helicopter crews, with dead and wounded, were still trapped on the battlefield and the ARVN attack was at a standstill. The Viet Cong even attempted, unsuccessfully, to destroy the two H-21's that remained sitting forlornly in the fourth lift's landing zone with mortar fire and rocket propelled grenades. The only encouraging sight to Lt. Col Vann as he orbited the battlefield, was Cpt. Ba's M-113's moving up on line. But, in a bitter disappointment, the M-113's fared no better than the armed helicopters, for the Viet Cong concentrated their fire on the carrier's exposed gunners, causing such casualties that the M-113's backed off. Each time they advanced, heavy fire and casualties forced them back. The helicopter crews from the 93rd and the UTT were left abandoned on the battlefield for at least seven or eight hours before finally being rescued.[27]

The enemy positions were further pounded by artillery, USAF 'Farm Gate' B-26's, and VNAF T-28's and A-1's, to no apparent effect. Fearful of the bad publicity and President Diem's ire that this disastrously unfolding day would bring, General Huynh Van Cao, who had been elevated to command the newly created IV Corps, flew into Tan Hiep at midday from his headquarters in Can Tho. In order to salvage something from the field of battle, General Cao determined to do whatever he could to bring this unhappy episode to an end. He requested a parachute drop, but instead of dropping the battalion of paratroops to the east of Ap Bac, where Lt. Col. Vann wanted them, to cut off a possible enemy escape route, General Cao intended to place the drop to the west so that they could reinforce the other ARVN units in the field; thus he was hoping to encourage the Viet Cong to disengage by giving them a safe route to withdraw. In a final act, adding insult to injury, the paratroops were actually dropped too late in the day, and much closer to Ap Bac than General Cao had ever intended, and the Viet Cong inflicted substantial casualties on the paratroops as they attempted to regroup in the growing dusk. By the end of the day, the field was left to the enemy. During the night, the Viet Cong slipped away, leaving behind snipers who continued to harass aircraft recovery teams from the 93rd Transportation Company's (Lt Hel) 80th Cargo Helicopter Field Maintenance (CHFM) Detachment, the 45th Transportation Battalion and the 611th Transportation Company (Direct Support).[28]

A camouflaged ARVN M-113 armored personnel carrier, photographed in January 1963, illustrates the exposed .50 caliber gunner's position which was the cause of so many casualties to Captain Ba's unit when he finally assaulted the enemy's prepared fortifications at Ap Bac on 2 January 1963. US Army Photograph

After Ap Bac both sides were forced to realistically evaluate the changed conditions. The Americans questioned their own tactics, ARVN competence and the limitations of the rules of engagement. General Harkins flew down to Tan Hiep on 3 January 1963 to get a first hand account of the battle from Lt. Col. Vann. One of Lt. Col. Vann's complaints was that the helicopter companies had standard operating procedures to send aircraft in to rescue downed crew. In Lt. Col. Vann's opinion, often quoted in contemporary press reports as an anonymous senior officer or top American commander, the helicopter pilots made a tactical error attempting to rescue the crews of downed helicopters. He saw it as misplaced chivalry, and frankly felt that the crews were, first and foremost, soldiers, no different from the U.S. Army advisors that rode in the same helicopters, off-loaded with the ARVN unit that they advised and were expected to remain on the field of battle and fight. Army aviators generally rejected this philosophy, for they were members of a tight knit fraternity that would routinely and unhesitatingly risk their lives to save one of their own. As far as Army aviators were concerned though, one fact had been made brutally clear at Ap Bac, and that was the realization that the nature of U.S. involvement was changing from one of being strictly in support of the ARVN to one of direct combat participation. As a means to express frustration with the new realities, and the sometimes bitter resentment of losing friends to the enemy, and partially to 'sound off' against MAAG, MACV, Saigon, the ARVN and all the rules they had to adhere to, a member of the 93rd dashed off a few satirical verses, sung to the tune of "On Top of Old Smokey". An abreviated version of this song was published in an April 1963 Newsweek Magazine article headlined 'South Vietnam: On Top of Old Ap Bac'.[29] Readjusting after Ap Bac was not made any easier by press reports of a decisive ARVN defeat, particularly since the U.S. Command had called Ap Bac a victory for Saigon. So disheartening were the reports in English language publications, with critical and unflattering descriptions of the ARVN's performance, that General Harkins returned to Tan Hiep on 4 January 1963 wanting to relieve Lt. Col. Vann on the spot. Bad news, and criticism of the ARVN, was not to be aired in public!

For the enemy the stand at Ap Bac also had extraordinary significance. The Viet Cong had not only engaged a vastly superior ARVN force but they had gained a victory. They had proven to themselves that they could counter the ARVN's increased mobility and that the awesome American war machines were not invincible. More ominously the Viet Cong could be expected to become increasingly aggressive, fielding larger caliber anti-aircraft weapons and refining their tactics to neutralize the advantages of airmobility. To commemorate the triumph, North Vietnam issued a supposedly 'National Liberation Front' stamp depicting the 'victory' at Ap Bac.[30] Two years later, in 1965, a small book was published in Hanoi, in English, titled 'Ap Bac - Major Battles in South Vietnam 1963 - 1964.'[31]

The battle of Ap Bac had a significant, long term impact upon Army aviation's role in the Vietnam War, from finally convincing the U.S. Army Support Group command to relocate the 57th Medical Detachment (Hel Amb) to Tan Son Nhut to rethinking the rules of engagement. The events that had transpired at Ap Bac were major contributing factors in the increased pressure on MACV, and the South Vietnamese government, to relax the restrictions imposed upon Army aviators. The nature of U.S. involvement had changed. No longer would Army aviators provide a simple helicopter transport service for the ARVN, as first envisioned by General Taylor and later recommended to President Kennedy in 1961. Instead the Army aviator's role was evolving into a more complex and lethal one; that of a full fledged combatant, who could actively seek out, hunt down and kill the enemy. It is instructive to note that it was in early 1963 that the new airmobile light helicopter companies were constituted with a gunship platoon and that combat arms officers, who were also rated Army aviators, were selected for command of aviation units, exemplified by Lt. Col Kenneth D. Mertel, Infantry, who was the first combat arms officer to assumed command of the 45th Transportation Battalion in August of 1963.[32] In February 1963 the rules of engagement, which previously limited defensive fire only in response to being fired upon, were changed to allow "the armed helicopter to initiate fire against clearly identified insurgents who threatened their safety or the safety of escorted transport helicopters." [33]

The UTT, who had discovered at Ap Bac "that the effectiveness of helicopters against prepared positions was limited" [34] continued to expand their role, adding reconnaissance and convoy protection to their usual duty of escorting the transport helicopter units. In February 1963 the 1st Platoon of the UTT was deployed to Da Nang to support the Marines. The pilots of the Marine helicopter squadron quickly became believers in the armed helicopter concept, and integrated the Army's armed escorts into their flight operations. The 1st Platoon was to remain in I Corps for almost the entire year, only rejoining the remainder of the company in December 1963. In recognition of its pioneering role, its major accomplishments and distinguished record the UTT was chosen to receive the Hughes Tool Company's, Aircraft Division, annual 1963 "Outstanding Army Aviation Unit" Award.

On 25 May 1963 four H-37B's of A Flight, 1st Platoon, 19th Transportation Company left Inchon harbor, Korea on the USNS Brostrom on their way to Vietnam. The unit, under the operational control of the Army Concept Team In Vietnam, arrived in Saigon on 3 June 1963 and the aircraft were split up, two supporting the 339th and two the 611th Transportation Companies (DS). A Flight, 1st Platoon, 19th Transportation Company was attached to the 611th Transportation Company (DS) for rations and quarters, but this attachment was later changed to the 330th Transportation Company (GS) also in Vung Tau. The four H-37B's, capable of recovering downed aircraft without major disassembly, provide a much needed heavy lift capability to the two Direct Support Companies.

The arrival of a new aviation unit in March of 1963 heralded the beginning of the end for the H-21's. Stationed in Vinh Long, the 114th Aviation Company (Air Mobile Light) was equipped with the Bell UH-1B. While the UH-1B Huey was destined to replace the

AP BAC * AFTERMATH

The following photographs, many of them rare and historic, of the fourth lift's landing zone at Ap Bac the day after the battle, and of the subsequent recovery efforts by the 611th Transportation Company (DS) and others, were taken while under sporadic enemy sniper fire. Most of these photographs, taken on 3 and 4 January 1963, are presented for the first time, the majority of which were taken by CW4 Frank Baldwin, USA Ret. formerly of the 93rd Transportation Company (Lt Hel) and Col. Sam Conley, Jr., USA Ret. former commander of the 611th Transportation Company (DS).

A view of the 'killing zone' from the enemy's fortified tree line.

The excellent fields of fire available to the Viet Cong are readily apparent in this photograph taken from on top of the enemy occupied dike, looking out over the rice paddies towards the downed aircraft. Of note is the small, half-submerged sampan loaded with rice stalks. These craft were fashioned by the villagers from logs that they had burned and carved out, and were used extensively by the Viet Cong to move reinforcements back and forth during the battle.

Both H-21C's of the 93rd and the UH-1B of the UTT positioned where they fell on 2 January 1963. The enemy lay in wait in the tree line in the background.

'Chalk 5' H-21C, s/n 51-15908, aircraft 'Y' and 'Chalk 6' H-21C, s/n 51-15903, aircraft 'D' of the 93rd Transportation Company (Lt Hel) and the UH-1B of the UTT receive the attention of recovery teams. Aircraft 'Y' still carries blue flight platoon markings on its nose and its aircraft letter in white on the forward rotor pylon; the conspicuity stripe on the fuselage top is red.

Unusual view of the downed UH-1B, s/n 62-1882, of the UTT as it awaits recovery. The violence of the crash is readily apparent. The aircraft was painted in gloss olive drab with high visibility markings.

H-21C, s/n 51-15908, aircraft 'Y' had replacement engine covers installed on both sides of the aircraft, each of which still carried a portion of the National Star Insignia.

All photographs on this page courtesy of CW4 Frank Baldwin, USA Ret.

*** NLF stamp issued in late 1963. More propaganda label than actual postage stamp, this is one of the rarest of so-called 'Viet-Cong' stamps.**

It would appear that this particular aircraft of the 93rd received its baptism of fire and first enemy hit in late March 1962 in I Corps. Note the personal marking 'DROOPY','awarded' the Purple Heart on 29 March 1962, painted in shades of light gray, red and purple behind the co-pilot's window on H-21C, s/n 51-15903, aircraft 'D'.

'DROOPY' also appears painted behind the aircraft commander's cockpit window, favorite locations for personal markings in the 93rd.

All photographs on this page courtesy of CW4 Frank Baldwin, USA Ret.

Right: **'DROOPY', H-21C, s/n 51-15903, aircraft 'D' in profile. The yellow letter 'D' painted on the removed front rotor pylon cover can just be seen stowed inside the cargo door. The aircraft did not carry any flight platoon colors.**

Below right: **H-21C, s/n 51-15903, aircraft 'D' of the 93rd Transportation Company (Lt Hel) showing to advantage the reduced size aircraft letter in yellow on the tail cone and the weathered, red conspicuity stripe on the fuselage top. The individual on top of the aircraft is believed to be SFC Ivon Coleman, Recovery Team Chief of the 611th Transportation Company (DS). The markings on H-21C aircraft 'Y' can also be readily seen.**

Below Left: **H-21C, s/n 51-15903, aircraft 'D', 'DROOPY' of the 93rd Transportation Company (Lt Hel) as it appeared at Ap Bac on 2 January 1963. This aircraft, 'Chalk 6', flown by CWO's Carl Flippin and Carleton J. 'Joe' Nysewander, was shot down during the action and later recovered successfully on 4 January. The aircraft carries the reduced size yellow aircraft letter 'D', personal markings and a somewhat muted and faded red conspicuity stripe.**

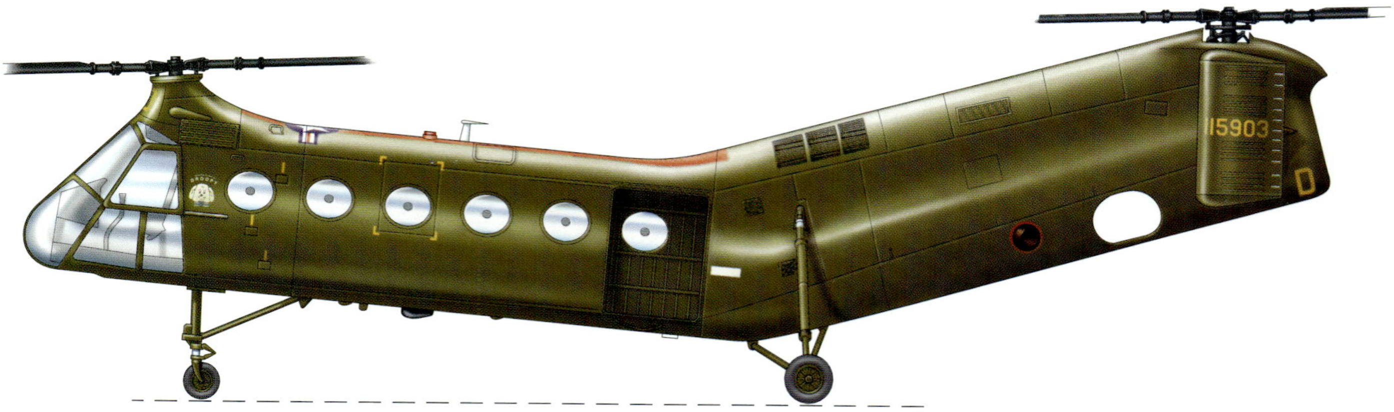

An appreciation of the knee-deep water and mud, which slowed each ARVN attack, can be gained from this photograph of a recovery team member heading towards H-21C aircraft 'Y'.

Right: Two maintenance men of the 80th Transportation Detachment (CHFM) inspect aircraft 'D' of the 93rd. The accurate enemy fire directed at the position of the crew chief, SP4 Donald Braman, can be readily seen.

US Army Photograph

Right: As recovery efforts on aircraft 'Y' are ongoing, a H-21 lands in the background.

Below: UH-1B, s/n 62-1883, of the UTT prepares to recover its own after the battle. It is of interest to note that subsequent unit histories do not mention losing any UH-1B's at Ap Bac, which has led to this photograph being misinterpreted by many historians. There is a widely held misconception that the UTT aircraft downed at Ap Bac was a UH-1A.

US Army Photograph

Above: Although not the best quality, this unique and rare photograph shows H-21C, s/n 51-15908, aircraft 'Y', in the middle foreground, on 4 January 1963 right after Captain Strever attempted lift-off, hit a rice paddy levee and rolled the aircraft. To facilitate recovery the 611th Transportation Company (DS) chopped the aircraft in half at the field splice, thereby earning the sobriquet 'Axes.'

All photographs on this page, unless otherwise noted, are courtesy of Col. Sam Conley, Jr., USA Ret.

Right: H-21C, s/n 51-15908, aircraft 'Y' being prepared for recovery by the 611th Transportation Company (DS). The general lack of cover and the wet conditions are shown to advantage in this photograph.

H-21's in Army helicopter units it must be kept in mind that the H-21 was not phased out all at once; it soldiered on in Vietnam until mid 1964. For one H-21 unit, however, it appeared that Nature itself was determined to speed up the transition process. 11 June 1963 came to be known as 'Black Tuesday' to the members of the 57th Transportation Company (Lt Hel). Earlier in the day, nine H-21's of the 57th Transportation Company (Lt Hel) participated in a large scale, multi-unit operation and soon after the flight was released to return to Tan Son Nhut, their home station, a monsoon season thunderstorm approached from the west and it began to rain. "Suddenly a violent, tornado-strength wind developed in the vicinity of the 57th's flight line and Operations building. Before anything could be done, estimated 80 MPH winds had swept down the flight line and overturned several of the aircraft. Although the wind lasted for only about 5-10 minutes, it left 12 H-21's and 4 TO-1D's damaged in its wake. The 57th was out of the flying business." [35] It was a 'Black Tuesday' indeed and not the auspicious beginning that any newly formed unit would choose, for, twelve days later on 23 June 1963, the H-21 helicopter units underwent redesignation:

The 8th Transportation Company (Light Helicopter) stationed at Qui Nhon became the 117th Aviation Company (Air Mobile Light).

The 33rd Transportation Company (Light Helicopter) stationed at Bien Hoa became the 118th Aviation Company (Air Mobile Light).

The 81st Transportation Company (Light Helicopter) stationed at Pleiku became the 119th Aviation Company (Air Mobile Light).

The 57th Transportation Company (Light Helicopter) stationed at Tan Son Nhut became the 120th Aviation Company (Air Mobile Light).

The 93rd Transportation Company (Light Helicopter) stationed at Soc Trang became the 121st Aviation Company (Air Mobile Light).

The 45th Transportation Battalion (Transportation Aircraft) was itself to undergo redesignation on 24 September 1963, becoming the 145th Aviation Battalion.[36] As the number of Army aviation units increased in 1963, straining the resources of the 45th Transportation Battalion, a second command unit, the 52nd Aviation Battalion was deployed to Pleiku in March 1963 to assume control of all aviation units in I and II Corps, units previously under the control of the 45th Transportation Battalion.

As the UH-1B became available and was introduced into the Army inventory, few could have predicted that the Huey, in all its various configurations and models, would become the enduring symbol of the Vietnam War.

Aerial view of Soc Trang, home base to the 93rd Trans. Co. (Lt Hel) 'Soc Trang Tigers', on 8 February 1963. 15 H-21's are visible; H-21C, s/n 56-2056, and one other in the air, three just touching down, six on the apron, three in front of the 80th Trans. Det.'s (CHFM) hangar and maintenance area and the remains of an aircraft at the edge of the apron. Also visible are six UH-1's of the UTT, a U-1A Otter and an O-1 Bird Dog.

US Army Photograph

SECTION III • Chronology of Army Aviation in Vietnam 1961-Mid 1963

This chronology is adapted from an unofficial document published by the 8th Military History Detachment, 1st Aviation Brigade in May 1968. It indicates major events, arrival or in-country activation of units, initial locations and parent organization of assignment. Some corrections and additions have been made by the author, based upon published primary documentation, such as unit histories.

Arrival or Activation Date	Unit and/or Event	Major Unit of Assignment	Station
Oct 54	Avn Div and Flight Det activated	MAAG, Indochina	Saigon
Aug 61	YAC-1, s/n 57-3080, arrived in RVN	ARPA	
Oct 61	18th AOD arrived in RVN	USARYIS	Soc Trang
Nov 61*	USAF 'Farm Gate' Det arrived RVN [1]	MAAG	Bien Hoa
Dec 61	8th Trans Co (Lt Hel) arrived in RVN	USARYIS	Qui Nhon
Dec 61	57th Trans Co (Lt Hel) arrived in RVN	USARYIS	Tan Son Nhut
Jan 62	93rd Trans Co (Lt Hel) arrived in RVN	USARYIS	Da Nang
Jan 62	45th Trans Bn (Trans Acft) arrived RVN	USARYIS	Tan Son Nhut
Feb 62	339th Trans Co (DS) arrived in RVN	USARYIS	Nha Trang
Feb 62	18th Avn Co (U-1A) arrived in RVN	USARYIS	Nha Trang
Feb 62	MACV activated in RVN	CINCPAC	Saigon
Feb 62	MACV-MAAG Flt Det activated	MACV-MAAG	Tan Son Nhut
Mar 62	3rd RRU utilizes ARDF aircraft in RVN	USASA	Tan Son Nhut
Mar 62	U.S. Army Spt Group, Vietnam activated	MACV	Tan Son Nhut
Apr 62	57th Med Det (Hel Amb) arrived in RVN	USASG-V	Nha Trang
Apr 62*	First USMC Helicopter unit arrived RVN [2]	MACV	Soc Trang
Jul 62	1st Platoon, 1st Avn Co (CV-2) arrived with 8 aircraft in RVN	USASG-V	
Jul 62	UH-1A Flight Section arrived in RVN	45th Trans Bn	Tan Son Nhut
Sep 62	ACTIV Avn Representative arrived in RVN		Saigon
Sep 62	23rd SWAD (OV-1) arrived in RVN	MACV	Nha Trang
Sep 62	33rd Trans Co (Lt Hel) arrived in RVN	45th Trans Bn	Bien Hoa
Sep 62	81st Trans Co (Lt Hel) arrived in RVN	45th Trans Bn	Pleiku
Sep 62	93rd Trans Co (Lt Hel) trades places with Marine helicopter squadron, HMM-163	45th Trans Bn	Soc Trang
Oct 62	UTT Helicopter Co arrived in RVN	45th Trans Bn	Tan Son Nhut
Oct 62	611th Trans Co (DS) arrived in RVN	USASG-V	Vung Tau
Oct 62	USASG-V auth one Avn Staff Officer each Corps Advisory Team auth 2 Avn Officers		
Nov 62	Strong staff action initiated by MAAG to relocate Avn Staff Section from MAAG to MACV		
Nov 62	UTT received UH-1B helicopters	45th Trans Bn	Tan Son Nhut

* Inserted in this chronology as a reference to the activities of other aviation elements mentioned in the text of Section II.

Arrival or Activation Date	Unit and/or Event	Major Unit of Assignment	Station
Dec 62	Avn Staff element consisting of 2 Officers activated at MACV		
Dec 62	Remainder of 1st Avn Co (CV-2) arrived RVN	USASG-V	Vung Tau
Dec 62	XM-6 Aircraft Weapon System introduced into RVN		
Jan 63	Battle of Ap Bac on 2 January - 93rd Trans Co (Lt Hel) and UTT involved in action		
Feb 63	Rules of Engagement modified		
Feb 63	1st Platoon, UTT sent to support USMC	MACV	Da Nang
Mar 63	114th Avn Co (AML) arrived in RVN	45th Trans Bn	Vinh Long
Mar 63	HHC, 52nd Avn Bn arrived in RVN	USASG-V	Pleiku
Apr 63	330th Trans Co (GS) arrived in RVN	USASG-V	Vung Tau
May 63	73rd Avn Co (O-1D) arrived RVN	45th Trans Bn	Nha Trang
Jun 63	'A' Flight, 19th Trans Co (H-37) arrived RVN	ACTIV	Vung Tau
Jun 63	H-21 cargo helicopter units redesignated on 23 June '63:		
	8th Trans Co - 117th Avn Co (AML)	52nd Avn Bn	Qui Nhon
	81st Trans Co - 119th Avn Co (AML)	52nd Avn Bn	Pleiku
	33rd Trans Co - 118th Avn Co (AML)	45th Trans Bn	Bien Hoa
	57th Trans Co - 120th Avn Co (AML)	45th Trans Bn	Tan Son Nhut
	93rd Trans Co - 121st Avn Co (AML)	45th Trans Bn	Soc Trang
Jun 63	Aviation Section, 3rd RRU organized in-country with RU-6A's and RU-8D's	USASA	Tan Son Nhut
Jul 63	Delta Avn Bn (Prov) organized in-country	USASG-V	Can Tho
Jul 63	61st Avn Co (CV-2) arrived in RVN	USASG-V	Vung Tau
Sep 63	45th Trans Bn redesignated 145th Avn Bn	USASG-V	Tan Son Nhut
Dec 63	1st Avn Co (CV-2) re-deploys to United States		

UH-1B's in the hold of the USNS Brostrom after arrival in the Port of Saigon, Vietnam on 9 March 1963. Serial numbers visible are 62-1888; 62-1889 [to the 57th Medical Detachment (Hel Amb)]; 62-1898; 62-1893 [UH-1 Flight Section, 45th Transportation Battalion, later redesignated 145th Airlift Platoon]; 62-1890 and 62-1892

US Army Photograph

SECTION IV • Painting, Camouflage and Markings of Army Aircraft 1961-Mid 1963

In any discussion of the painting, camouflage and marking of Army aircraft two documents are crucial; the first is the Department of the Army Technical Bulletin TB 746-93-2 subtitled Painting and Marking of Army Aircraft and the second is the Federal Standard No. 595a subtitled Colors. TB 746-93-2, as it relates to the standard authorized paint scheme of each Army aircraft type, will be dealt with in greater detail in the text that follows. Federal Standard (FS) No. 595a is organized into three sections, Volume I consists of text and 1/2 by 1 inch color chips, Volume II covers the spectrophotometric curves and tristimulus values (using an instrument known as a spectrophotometer which measures the relative intensity of light in different parts of the spectrum and is expressed as a plotted graph of percentage reflectance and wavelength in millimicrons) to be used in color control and reissuance of color chips and, lastly the third section, the Chips, which are individual 3 by 5 inch color chips (438 different color chips were available in the set issued on 2 January 1968).

Color - Universal Language and Dictionary of Names, by Kenneth L. Kelly and published by the Department of Commerce, National Bureau of Standards - NBS Special Publication 440, December 1976 distinguishes colors by hue, value and chroma. Basically hue refers to the primary color groups red, yellow, purple, blue and green; value describes the color's lightness to darkness as expressed in light green as opposed to dark green; and chroma indicates the amount of gray (called saturation) within the color. There are, literally, thousands of shades of color that can be distinguished by the human eye, and many variables exist in each individual's direct observation of a color. The U.S. Army sought to standardize the paint colors authorized to be applied to each Army aircraft by referencing colors identified by a 5-digit numbering system in FS No. 595a. Colors were classified into color groups, by hue, and color classifications by specular gloss, designated as full gloss, semigloss and lusterless. In general usage these three categories are referred to as gloss, semigloss and flat. In FS No. 595a the 5-digit numbering system is broken down as follows:

The first digit indicates finish, 1- full gloss, 2 - semigloss and 3 - flat;

The second digit indicates the color groups by predominant hues;

The last three digits indicate the order of increasing reflectance, or diffusion, listed in order of darker to lighter values.

For example, in the text, the color expressed as Gloss Olive Drab (FS. 14087) is 1 - Gloss, 4 - in the predominantly green group, and 087 somewhat darker in intensity than most greens. Conversely FS. 34087 is Flat Olive Drab. It should be noted that in the early 1960's the Army was using an interim color chip numbering system, the prefix "X" being used as in Olive Drab X34087 (Lusterless). The Director of the Department of the Army Mobility Equipment Research & Development Center, Coating and Chemical Laboratory wrote, in a letter to the author, "Since Federal Standard No. 595a was issued the interim chips with the "X" prefix are no longer used."[1]

The material that follows, covering the painting, marking and camouflage of each major Army aircraft type in service in Vietnam during the time period covered, late 1961 to mid 1963, references both TB 746-93-2 and Federal Standard No. 595a; the interim color chip numbering system, with the "X" prefix, will not be referenced in the text.

It should be kept in mind, while studying photographs of Army aircraft in Vietnam, that there are many factors which will impact an aircraft's appearance, such as the weathering action of the sun, wind, rain, saltwater and dirt, paint deterioration and discoloration due to age, paint application from minor unit field touch-ups, due to combat damage, to major direct and general support maintenance activities repainting, command required markings, unit required insignia and markings, personal markings and even human error in paint selection and marking application.

VERTOL AIRCRAFT CORPORATION H-21 SHAWNEE

After the Kennedy administration's decision was made to send Army aviation units to assist South Vietnamese forces in late 1961, the H-21 cargo helicopter units were alerted and deployed too quickly for any serious thought to have been given to the desirability of a new paint scheme, or even camouflage, for Southeast Asia. In fact, since U.S. helicopter units were intended to be strictly in support of the ARVN, if any thought was given to the problem it was probably considered advisable to keep the standard, authorized paint scheme in use at that time, without modification. The Kennedy administration was aware that President Diem, of the Republic of Vietnam, had been so uncomfortable with the idea of U.S. forces in his country that he had floated the idea that U.S. civilian pilots should be sent to fly helicopters and C-47 transport aircraft in Vietnam as late as October 1961.[2]

However the five H-21 cargo helicopter units were not always uniform in appearance when they landed in South Vietnam, since there were stringent requirements levied on these units prior to deploying, and they often had to obtain replacement aircraft from other units in order to be at full strength. The 93rd Transportation Company (Lt Hel) had even replaced one of its aircraft's engines, while in transit through the Philippines, with an already 'built up' replacement engine flown in by the USAF![3]

Army aircraft were delivered by the manufacturer in an overall paint scheme according to the Army standards in force at the time of delivery. These standards were published in an Army Technical Bulletin, TB 746-93-2, which was updated every few years to account for the new aircraft types coming into the Army's inventory. TB 746-93-2 clearly spells out, in both drawings and text, the authorized paint scheme for each aircraft type, such as the H-21.

The authorized paint scheme for the H-21 was overall Gloss Olive Drab (FS. 14087). The 24 inch high 'ARMY' lettering on the forward fuselage was Gloss White (FS. 17875) and the legend, in 6 inch high lettering, 'U.S. ARMY' and the 8 inch high aircraft radio call numbers, derived from the ser-

H-21C, s/n #55-4171, of the 57th Transportation Company (Lt Hel) at Son Dong airstrip on 2 September 1962. Aircraft is painted in overall gloss olive drab, with the Army's standard high visibility markings, and carries all regulation stenciled data. The 57th added the 'last two' of the serial number in gloss yellow to the forward rotor pylon.

U.S. Army Photograph

Figure 5: Vertol H-21 composite illustration showing the placement of conspicuity, unit and personal markings carried by the cargo helicopter companies in Vietnam 1961 to mid 1963.
Illustration by Steve Davis

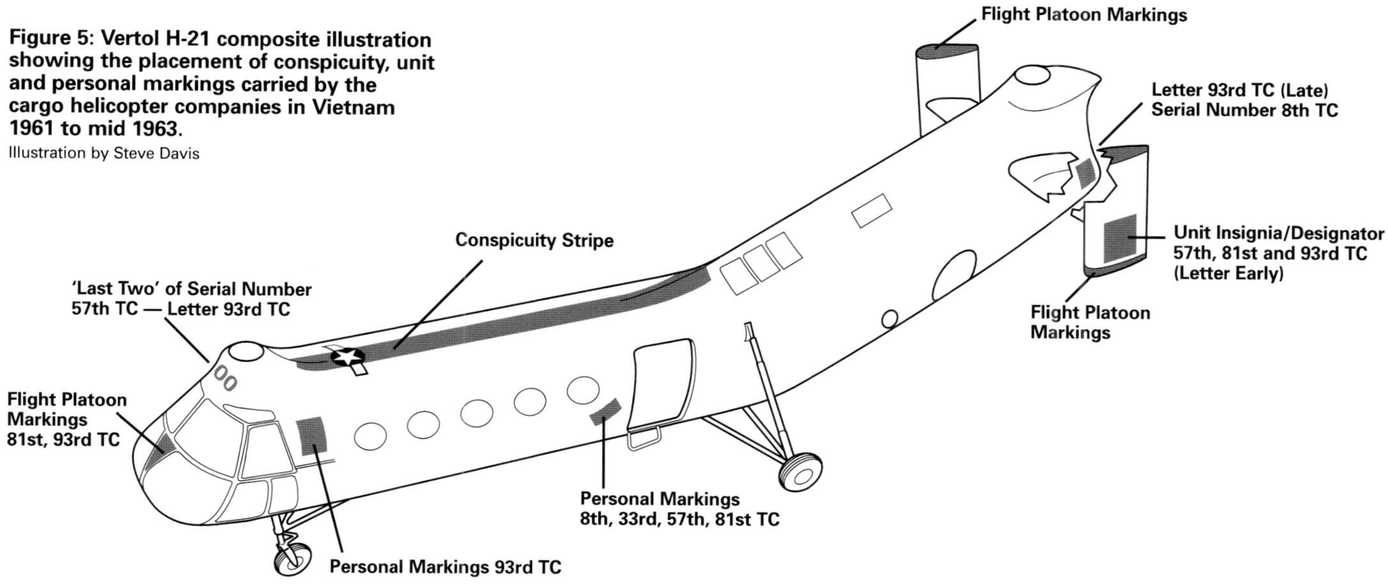

(Note: all other markings, such as aircraft serial number, left off for clarity)

ial number, on the vertical tail fin surface were painted in Gloss Yellow (FS. 13538). The Technical Bulletin called for four National Star Insignia to be carried by rotary wing aircraft. In the case of the H-21 two 50 inch diameter National Star Insignia were carried on the rear fuselage, on a line parallel to the cargo floor, aft of the main door. Two smaller 30 inch diameter National Star Insignia were positioned on the forward top and bottom of the fuselage. Multiple exterior maintenance data, escape panel markings, aircraft serial number, fuel specification etc. in yellow, white and black lettering could be found, positioned as indicated in the Technical Bulletin, at various places on the aircraft. Often the majority of this data did not survive an aircraft repainting, particularly if done in the field. Interiors were Gloss Gray (FS. 16440). Seat frames were painted in a Lusterless Gray (FS. 36440) while early aircraft seat cushions and nylon webbing were olive drab in color. Later arriving aircraft had the newer, red colored cushions and webbing. The rotor blades were painted flat Black (FS. 37038) on the undersides of the blade and light Gray, (FS. unknown) on top.

A stripe marking on top of the fuselage, initiated by the 93rd Transportation Company (Lt Hel) H-21's in Da Nang in mid 1962, had spread to the remaining cargo helicopter units by September 1962. This first theater-wide Army aviation marking was then standardized, in all probability, under the direction of the 45th Transportation Battalion, and generally consisted of a Gloss White (FS. 17875) high visibility stripe. The stripe was approximately 24 inches wide as applied to the fuselage top of the H-21 helicopter and a thinner, approximately 6 inch wide, stripe when applied to the fuselage top and wings of fixed wing aircraft such as the U-1A Otters of the 18th Aviation Company. The purpose of these white stripe markings was to make it easier to spot the aircraft from above, and to facilitate search, rescue and recovery operations for downed aircraft. Although white was the color most often used, and observed, some units appear to have interpreted the instructions in their own way and used colors other than white, even a two-color stripe, as carried by one H-21 unit.[4] A former member of the 93rd Transportation Company (Lt Hel), who often directed H-21 flights from the unit's TO-1D, commented on the use of the stripe marking as a means to locate lower flying aircraft "....they were difficult to spot lower against the green foliage. I suspect the markings may have been put on for that. However you would pick the flights up visually by sun reflecting off rotors...." [5] These markings would come to be officially called 'conspicuity' markings by the mid-1960's.

By September of 1962 some H-21 units had begun to paint over both the 50 inch on the fuselage and the 30 inch National Star Insignia on the belly of the aircraft, leaving only the 30 inch National Star Insignia on the fuselage top. Shortly thereafter, the 24 inch high white 'ARMY' on the fuselage was painted out. These markings were painted over with whatever stocks of olive drab paint were on hand, or whatever was locally available; often the overpainting being done in a Flat Olive Drab (FS. 34087) rather than a gloss paint, resulting in rather patchy looking aircraft. This was done to prevent presenting enemy gunners with a highly visible aiming and tracking point. There does not seem to have been any consistent implementation, by the 45th Transportation Battalion for example, of these first attempts at toning down the aircraft markings; it appears to have been left to the discretion of each unit.

The H-21 units had arrived in Vietnam displaying a variety of unit markings, and in one case, the 81st Transportation Company (Lt Hel), unit insignia on their aircraft. The unit markings underwent evolution for the same reason that the unit's toned down their aircraft markings, the less conspicuous an aircraft was, the better.

Most H-21 units allowed colored flight platoon markings usually in white, blue, yellow or red. Flight platoon markings were carried as trim on the vertical tail, top and bottom fairings. After the toning down of markings in late 1962, flight platoon colors were generally confined to the lower fairing of the tail surface. Often flight platoon colors were also carried on the nose triangle created by the cockpit framing, particularly in the 81st and 93rd Trans. Co.'s. One unit, the 57th Trans. Co., repeated the last two digits of the aircraft's serial number in gloss yellow on the forward rotor pylon. In rare cases, where more than one aircraft within the unit shared identical last two serial numbers, a third digit was added to the forward rotor pylon to distinguish the two aircraft. [6]

At the discretion of the unit commander, aircrews added personal markings to their aircraft, most frequently in the form of a name, but artwork, to accompany the name, was becoming increasingly common by the first half of 1963. There was a great deal of consistency, within most units, in the positioning of the personal markings, most being placed just forward of the main entrance door, but the 93rd Transportation Company (Lt Hel) was unique, often placing personal markings behind the pilot and co-pilot's sliding cockpit windows.

First observed in late 1962, when the 8th Transportation Company (Lt Hel) camouflage painted their helicopters, and spreading to the 57th Transportation Company (Lt Hel) in March 1963, unique, non-standard camouflage schemes evolved in two H-21 units. In late 1962 and early 1963 the aircraft appear to have been painted in a three-color camouflage scheme consisting of tan and a light blue-green applied over the aircraft's dark, base-coat of Gloss Olive Drab (FS. 14087). Since these schemes appeared to have been most often hand painted in the field, at least in the 57th Transportation Company (Lt Hel), there was no apparent attempt to develop a standard camouflage pattern to guide application. Each aircraft reflected a different design, application being neat and precise or crude and patchy, depending on the aptitude, or lack thereof, of each painter. The colors seem consistent in the photographs, however it has been hard to determine the exact Federal Standard (FS) color used, if indeed there was any, for the source of the paint may not have been U.S. military stock, but local acquisitions.

It should be noted that these early, 1962-1963, H-21 camouflage schemes appear to be distinct from the camouflage schemes in use on H-21's in the first six months of 1964 by the successor units to the Transportation Companies. It would seem that there were two distinct H-21 camouflage scheme phases; the first early phase, December 1962 to late 1963, as described above, where the three colors used appear darker on the aircraft, and a unique, and very distinct, blue-green color appears. The second, later phase, roughly between November 1963 to June 1964, where, in general, two camouflage colors were used, large patches of tan applied over the darker olive drab paint, with tan predominating.

'Chattanooga Choo-Choo', an H-21C, s/n unknown, of the 57th Transportation Company (Lt Hel) on-loads ARVN troops at Ca Mau airstrip. After September 1962 the high visibility ARMY and National Star Insignia on the fuselage were 'toned down' by being over-painted in various shades of olive drab as seen here. US Army Photograph

UH-1B, s/n 62-1881, being unloaded from an USAF C-133 transport at Tan Son Nhut on 29 November 1962. This UTT aircraft is painted in overall Gloss Olive Drab (FS. 14087), and carries regulation National Star Insignia and high visibility markings.
US Army Photograph

UH-1A, s/n believed to be 59-1695, of the UTT at Tan Son Nhut on 21 February 1963. Except for the US ARMY and serial number in yellow on the tail fin, and roof and belly National Star Insignia, the aircraft has been over-painted in gloss olive drab. SP5 Weldon B. Reynolds, left, and SP5 Joseph E. McGurk load rockets into the MA-2 rocket launcher.
US Army Photograph

BELL UH-1A AND UH-1B IROQUOIS (HUEY)

The UH-1A's of the Utility Tactical Transport Helicopter Company (UTT), while working up for combat in Thailand and as they first arrived in Vietnam, carried the Army's standard high visibility paint scheme, and carried the colors and markings of the three respective flight platoons. Based upon the Technical Bulletin both the UH-1A, and the UH-1B, were painted in overall Gloss Olive Drab (FS. 14087) with a 30 inch wide Gloss Yellow (FS. 13538) band around the tail boom behind the synchronized elevator, with a red arrow and the wording DANGER above and KEEP AWAY below the arrow centered in the tail band. A 24 inch diameter National Star Insignia was applied to either side of the tail boom and both on top of the cabin and on the bottom of the fuselage. A Gloss White (FS. 17875) ARMY legend in 30 inch high letters was positioned behind the National Star Insignia on the tail boom. A Flat Black (FS. 37038) anti-glare panel covered the nose and the instrument shade panel in the cockpit. The aircraft interior was Gloss Gray (FS. 16440) and the seat cushions and nylon webbing was red. The aircraft's radio call numbers/serial numbers were painted in Gloss Yellow (FS. 13538) in six inch high letters on the upper tail fin, and the yellow legend U.S. ARMY, also in six inch high lettering, was painted above the call numbers. Numerous maintenance, warning, and the aircraft's data block, fuel filler cap, etc. markings were applied in yellow, white, red and black as specified by the Technical Bulletin. The rotor blades were Flat Black (FS. 37038) on the undersides and Olive Drab (FS. 24201) on the top. The last six inches of the blade tips, on both surfaces, were painted in Flat Yellow (FS. 33538), while the tail rotors were painted in high visibility red, white and red three inch bands on the outer tip of the black blade, which had another three inch red band at the center.

The UH-1A's of the UTT also had their markings toned down, to reduce visibility, in or around late October, or early November 1962. They were re-sprayed Gloss Olive Drab (FS. 14087) covering the tail boom markings and yellow tail band, leaving the National Star Insignia on top of the cabin and the bottom of the fuselage, and the U.S. ARMY and serial number remaining on the tail fin. In late November 1962 the UTT was augmented by brand new UH-1B's, also delivered in the standard gloss olive drab high visibility markings. Until mid 1963 the remaining low visibility gloss olive drab painted UH-1A's operated alongside the high visibility painted UH-1B's. Eventually, in or around June 1963, the Army Concept Team In Vietnam (ACTIV) became involved in the evaluation of a three-color camouflage scheme on at least three of the UTT's UH-1B's.

FIXED WING AIRCRAFT - CESSNA L-19 (TO-1D) BIRD DOG AND GRUMMAN JOV-1C MOHAWK

The decision having been made to replace the cargo helicopter unit's organic H-13's with Bird Dogs, the aircraft were assigned to the H-21 units painted in standard high visibility markings, the same painting scheme

T0-1D, s/n 55-4676, of the 93rd Transportation Company (Lt Hel) is refueled at the Vinh Long airstrip on 21 February 1963. Aircraft is painted in Gloss Olive Drab (FS. 14087), carries the National Star Insignia and high visibility markings. US Army Photograph

that the Mohawks of the 23rd Special Warfare Aviation Detachment (SWAD) carried when they arrived in Vietnam.

The fixed wing aircraft, including the U-6A's and U-1A's, were painted overall Gloss Olive Drab (FS. 14087). Fixed wing aircraft carried the National Star Insignia on both sides of the fuselage and on top of the left wing and underneath the right wing. The L-19 Bird Dog carried 30 inch diameter National Star Insignia on the wings, and 20 inch diameter National Star Insignia on the fuselage. The lettering, 18 inch high ARMY in white on the fuselage, and 18 inch high U.S. ARMY on both the top right wing and under the left wing, and the 7-1/2 inch high U.S. ARMY and 12 inch radio call numbers on the tail were in Gloss Yellow (FS. 13538). A Flat Black (FS. 37038) anti-glare panel was painted on the nose.

The Mohawks of the 23rd SWAD carried 30 inch diameter wing and fuselage National Star Insignia. The ARMY marking was carried in 18 inch high gloss white letters on the fuselage and the U.S. ARMY marking was carried in 20 inch tall letters on the top of the right wing and the bottom of the left wing in gloss yellow. U.S. ARMY lettering and the radio call sign (serial numbers) were carried in 8 inch high gloss yellow letters on the outer vertical tail fin surface. Flat black anti-glare panels were painted on the nose and the inside of the engine nacelles. Numerous maintenance, warning, danger and ejection seat signs constituted the exterior data markings of the Mohawk.

In early Vietnam service with the 23rd SWAD the JOV-1C Mohawks carried Gloss Yellow (FS. 13538) aircraft identification numbers, approximately 24 inches high painted on the outer tail fin, and a smaller version, 12 inches high, positioned low on the nose. The first six aircraft were numbered 1 through 6. Through attrition and the unit's receiving additional, as well as, replacement aircraft, by early 1964 higher numbers, such as 12 and 17 had started to appear.

MACV-MAAG FLIGHT DETACHMENT AIRCRAFT

The aircraft flown by the MACV-MAAG Flight Detachment were painted in a color scheme designated 'Arctic, Desert and Jungle Markings'. In essence this was a peacetime paint scheme designed to improve an aircrew's post crash survival, in the jungle for example, by ensuring, hopefully, that the brightly painted red and white aircraft would be quickly located. The U-6A's and L-23's (U-8D) of the Flight Detachment were painted in a Gloss Red (FS. 11136) and Gloss White (FS. 17875) pattern. The gloss red portion on the wings, covering approximately half the wing span, did not cover the National Star Insignia and the Gloss Black (FS. 17038) U.S. ARMY lettering which were on the white background color cut-outs. Black ARMY lettering and the National Star Insignia were carried on the fuselage, while the tail was painted gloss red. The radio call numbers, or serial numbers, and U.S. ARMY were carried on the tail within white background color cutouts. Flat Black (FS. 37038) was used on the aircraft cowlings, where necessary, and the nose on the L-23, to cut down glare into the cockpit.

Photographs of the MACV-MAAG Flight Detachment aircraft show that American flag decals were carried on the unit's U-6A and L-23's through early 1964. The decals, approximately 17-1/2 by 9-3/4 inches, were positioned on the gloss red paint, above the white cutout for the serial number on the U-6A and below the cutout on the L-23 aircraft.

The L-23F being flown by the Detachment in early 1963 was painted in a gloss olive drab and white VIP scheme.

Note: The painting and markings carried by the Army's CV-2 Caribou and the U-1A Otter will be covered in Volume II - Army Aviation in Vietnam 1963 -1966.

SECTION V • Army Aviation Units in Vietnam 1961-Mid 1963

This section is intended to provide a 'snapshot' in time, to illustrate, through official U.S. Army photographs and personal photographs taken by Army aviators and unit members themselves, the atmosphere and character of the earliest days of Army aviation in Vietnam; with brief, unofficial unit histories, unit commanders and personalities and mention of unit traditions, mottoes, nicknames, accomplishments, and illustrations of unit facilities, where possible. It is further intended to illustrate the unit patch(es), and include photographs, where available, of the patch being worn on the uniform, and any other distinctive insignia that the unit may have worn, such as the 'beercan' insignia (so called because they appeared to have been made out of beercan metal), in order to show unit insignia in use on headgear, unit signs, operations buildings, hangars and on the unit's aircraft. Photographs, color profiles and drawings will illustrate the painting, camouflage and markings of a unit's aircraft and, where known, serial numbers will be identified in all aircraft photographs. The color profiles by Steve Davis and uniform studies by figure artist Kevin Lyles, specially commissioned for this volume, represent the most accurate color illustrations possible based upon extensive, in-depth research from original sources.

UNITS IN VIETNAM

45th Transportation Battalion (Transportation Aircraft)

18th Airfield Operating Detachment

57th Transportation Company (Light Helicopter) (H-21)

8th Transportation Company (Light Helicopter) (H-21)

93rd Transportation Company (Light Helicopter) (H-21)

33rd Transportation Company (Light Helicopter) (H-21)

81st Transportation Company (Light Helicopter) (H-21)

Utility Tactical Transport Helicopter Company

23rd Special Warfare Aviation Detachment

611th Transportation Company (Direct Support)

Aviation Section, 3rd Radio Research Unit

1st Aviation Company (CV-2)

MACV-MAAG Flight Detachment

57th Medical Detachment (Helicopter Ambulance)

A Flight, 1st Platoon, 19th Transportation Company (H-37)

UH-1A Flight Section, 45th Transportation Battalion (Trans. Acft.)

The following units, due to their many consecutive years of service in Vietnam and to avoid unnecessary duplication, will be covered in subsequent volumes:

18th Aviation Company (U-1A)

339th Transportation Company (Direct Support)

45th Transportation Battalion (Transportation Aircraft)

COMMANDER'S CALL SIGN "ALLIGATOR 6" 1963

Patch of the 45th Trans. Bn. (Trans Acft).
Patch courtesy of
Col. Kenneth D. Mertel, USA Ret.

VIETNAM SERVICE

The 45th Transportation Battalion deployed to Vietnam in January 1962 to assume control of the three helicopter companies in Vietnam, and served until 24 September 1963, when the unit was inactivated. All personnel, assets and equipment of the 45th were transferred to the successor unit, the 145th Aviation Battalion.

COMMANDERS, NOTABLE UNIT MEMBERS AND TRADITIONS

First Commander of the 45th was Lieutenant Colonel Howard B. Richardson who was responsible for setting up the necessary administrative and organizational structure which could effectively handle the influx of aviation units into Vietnam, including helping the Marine's helicopter unit, HMM-362, get settled in at Soc Trang in April 1962. The CO of the 45th, with his ever-present pipe, became a familiar figure to the units under his command. The 45th Transportation Battalion, under Lt. Col. Richardson, gained the distinction of being chosen the Hughes Tool Company, Aircraft Division's "Outstanding Army Aviation Unit" for 1962. The trophy was presented by the Army's Chief of Staff, General Earle G. Wheeler, who cited the 45th's Battalion Headquarters and its four attached Aviation Companies, the 8th, the 18th, the 57th and the 93rd, for their achievements, stating " these achievements include thousands of flying hours in transporting units of the Army of South Vietnam on combat missions. They include aerial resupply to besieged hamlets and the evacuation of wounded from the combat area. During many of these missions the helicopters of the battalion have been under groundfire. They have had pilots and crewmen wounded and killed. They have operated widely dispersed over a large geographic area and under conditions of weather and terrain that are punishing both to men and to machines."[1] The trophy was accepted on behalf of the 45th Transportation Battalion by Majors Milton P. Cherne and William J. Tedesco, Major Cherne delivering the acceptance speech.

In September 1962 Lieutenant Colonel Robert L. Hoffman assumed command, and he continued to build morale, a 'can do' attitude and professionalism within the battalion. He strove to improve the maintenance supply system for the H-21 units and often personally 'dropped in' to attend mission briefings at his units. Not everyone could, or did, live up to Lieutenant Colonel Hoffman's high expectations, and, consequently "The man

Below: **Lieutenant Colonel Howard Richardson, and his staff, Captains Peppard and Bass, flew down to attend Major Cherne's briefing of the newly arrived Marine helicopter squadron, HMM-362, at Soc Trang, April 1962. Visible individuals, left to right, standing next to the Otter's door Captain George Peppard, Captain Sam Bass, Major Milton Cherne, Lieutenant Colonel Blackmore with back to camera, USMC, unknown, Lieutenant Colonel Howard Richardson with pipe, and two unknowns, possibly the Otter's crew chief and crewman. The U-1A Otter, s/n 55-3272, may have been assigned to the 18th Aviation Company (U-1A).**

Photograph taken by and courtesy of
Col. Emmett F. Knight, USA Ret.

who led this unit was no doubt "damned" at times by the uninformed and new arrivals, but it was he who inspired them all to go on and accomplish what had to be done in the best possible manner. It was he who supervised, chewed and encouraged."[2] In mid 1963 the 45th was inspected by USARYIS, the inspection conducted to full stateside standards. The 45th passed with flying colors, all this while in the midst of continuing combat support to the ARVN. In August 1963 Lieutenant Colonel Kenneth D. Mertel assumed command, the last Commanding Officer of the 45th Transportation Battalion (Transportation Aircraft) before it was inactivated 24 September 1963, at which time he became the first Commander of the successor unit, the 145th Aviation Battalion.

Command emphasis on effective maintenance resulted in a fairly visible maintenance officer for the 45th Transportation Battalion. At the beginning of the year in 1963, the 45th's Maintenance Officer during that time, Captain Carl W. Strever was awarded the Distinguished Flying Cross for his courageous actions after the battle at Ap Bac, when he assisted in aircraft recovery operations under enemy sniper fire. The presentation of the DFC to Captain Strever was made by the Army's Chief of Staff General Wheeler during one of his visits to Vietnam. Captain Strever designed the tiger head emblem which became the insignia and patch of the 121st Aviation Company (AML) the famed "Soc Trang Tigers"[3] Captain Strever had also 'liberated' a somewhat war weary Grumman F8F-1 Bearcat, formerly belonging to the Vietnamese Air Force. He flew the Bearcat recreationally as his personal mount in early 1963, to the amazement of the H-21 units, until the authorities found out about it![4] The Bearcat probably ended up as a static 'gate guard' at a VNAF instillation.

FACILITIES

When the 45th Transportation Battalion arrived at Tan Son Nhut, the 57th Transportation Company (Lt Hel) already had it's tent city compound well established at the airbase. The 45th set up in the same compound, eventually building and occupying wooden 'hooches' and sharing a consolidated mess administered by the 57th Transportation Company.

AIRCRAFT PAINTING AND MARKINGS

The 45th Transportation Battalion was assigned a few aircraft for the use of the Battalion Commander and his staff, and may also have supported the U.S. Army Support Group, Vietnam Commander and his staff. At least one L-23D (U-8D), s/n 57-6068, is believed to have been assigned painted in the 'Arctic, Desert and Jungle Markings' color scheme of Gloss Red (FS. 11136) and Gloss White (FS. 17875) identical to the MACV - MAAG Flight Detachment scheme except that no American flag decal was carried on the tail.

The F8F-1 Bearcat, Bureau Number (BuNo) 95338, still carrying the unit markings of the 1st Fighter Squadron, VNAF which Captain Carl Strever, a former Marine pilot, flew out of Soc Trang in February and March 1963.

Photograph courtesy of CW4 Frank Baldwin, USA Ret.

Lieutenant Colonel Kenneth D. Mertel, Commanding and Command Sergeant Major Benjamin Knight. While Lieutenant Colonel Mertel is still wearing the 45th Transportation Battalion patch, this photograph was actually taken shortly after the redesignation of the Battalion as the 145th Aviation Battalion in September 1963.

Photograph courtesy of Col. Kenneth D. Mertel, USA Ret.

Headquarters of the 45th Transportation Battalion, with the CO's M-38A1 parked in front, in June 1963.

Photograph courtesy of Col. Robert L. Hoffman, USA Ret.

Believed to be the 18th AOD's first patch design, 1961-1963

Second patch design of the 18th AOD, 1963-1965.
Patch courtesy of Chris Miller.

18th Airfield Operation Detachment (AOD)

VIETNAM SERVICE

The 18th Airfield Operating Detachment was sent to Vietnam from Okinawa in October 1961, being based at Soc Trang. Unit expanded it's operations to include all Corps areas by 1965, when it was inactivated. Personnel, assets and equipment were transferred to the 125th Air Traffic Company (Air Traffic Control) in August 1965.

UNIT NOTES

The 18th Airfield Operating Detachment was originally based in Soc Trang, where the unit set up its equipment, and attempted to make the rundown, former Japanese Army Air Force base habitable. Tents were home to the 18th AOD and the unit had not progressed much further when the Marines of HMM-362 arrived in April of 1962. Eventually as the duties and responsibilities of the 18th AOD were expanded, the unit had operating locations at airfields throughout Vietnam, and accommodations ranged from good to miserable.

The 18th AOD was commanded by Captain Gerald C. Guettler in the late 1964 time frame.

OPERATIONS

The 18th Airfield Operating Detachment's mission was to provide air traffic control for Army aviation units, as well as other user Services such as the USAF and USMC. Equipment that the unit used varied from shelters outfitted for air traffic control to control towers at Army Airfields, Ground Control Approach Radar Systems, Radio Relay equipment etc. Soc Trang soon boasted a control tower (TWR), a Ground Control Approach Radar System (GCA) and a Non-Directional Beacon (NDB).

From this modest beginning the 18th AOD spread out and installed a control tower at Pleiku, and another at Qui Nhon which also had a GCA. The remainder of I and II Corps areas were covered with flight following stations (FFS) and Non-Directional Beacons (NDB) . At Tan Son Nhut a control tower facility was established, eventually to be known as "Saigon Center", and a Flight Operations Center (FOC). Vung Tau also received a control tower and a GCA. Flight Following Stations (FFS) and NDB's were located at other key air bases, from Phan Thiet to Ca Mau, to provide theater-wide coverage.

One navigational system appeared to work well for the H-21 units. The Decca navigation system was installed in several of a unit's aircraft, and following the installation of a third, slave station in the Delta, an effective navigational aid was created. The "Mark 817 Decca system consisted of a chart upon which the position of the aircraft was recorded by a pen. The pen was geared to a radio receiver which received pulses from prepositioned transmitters surrounding the area of operations." [1]

As Army aviation assets continued to build in the Republic of Vietnam, Major General Delk M. Oden, who had assumed command of the U.S. Army Support Command, Vietnam after General Stilwell in 1964, had requested more comprehensive Army Flight Following facilities, and the 18th's successor unit, the 125th Air Traffic Company (ATC) grew out of that request in 1965.

AIRCRAFT PAINTING AND MARKINGS

The 18th AOD was assigned two U-1A Otters, s/n's unknown, according to later unit histories. While no identified or published photographs of these two aircraft have been found, these Otters would most certainly have carried the gloss olive drab and standard high visibility markings current at the time. It is not known if unit markings or personal markings were carried by 18th AOD aircraft.

57th Transportation Company (Light Helicopter) (H-21)

FIRST UNIT NICKNAME "MILT'S MISFITS" 1962

SECOND UNIT NICKNAME "DEANS OF THE HELICOPTER CORPS" 1962-1963

98TH TRANSPORTATION DETACHMENT (CHFM)

VIETNAM SERVICE

December 1961 to 23 June 1963 when the unit was inactivated. All personnel, assets and equipment of the 57th were transferred to the successor unit, the 120th Aviation Company (Air Mobile Light) along with the attached 98th Transportation Detachment.

COMMANDERS, NOTABLE UNIT MEMBERS, TRADITIONS

First unit commander was Major Robert J. Dillard who successfully oversaw the deployment of the 57th to Vietnam, and shortly thereafter, upon promotion to Lieutenant Colonel, was transferred to MAAG, Vietnam. Major Milton P. Cherne assumed command, and soon after the arrival in Soc Trang of the Marine helicopter unit in April 1962, the 57th became known as "Milt's Misfits", in order to distinguish themselves from "Archie's Angels" of HMM-362.[2] Major Cherne later wrote a number of award-winning articles, for U.S. Army Aviation Digest among others, which assured the 57th Transportation Company (Lt Hel) its place in history. The third, and last, unit commander was Major Darwin D. Beauchamp. Major Beauchamp established a unit tradition in 1963 of a free drink to crewmembers of any aircraft that was hit by enemy fire. He also established an "Eagle Club and Eagle Aces" within the unit. Under Major Beauchamp's command the 57th was first helicopter unit to complete 10,000 flying hours in Vietnam on 1 April 1963.

The second nickname that the unit acquired had an interesting origin. Malcolm W. Browne, the Associated Press representative in Vietnam in the early 1960's, was well respected by the cargo helicopter units for his balanced, and generally accurate, reporting. He had written a feature story on the 57th in 1962 in which he called them the "deans of the helicopter corps" and the name struck a cord with the unit.[3] The nickname, eventually shortened to "the Deans" , became associated with the 57th, and stayed on with its successor unit, the 120th Aviation Company (Air Mobile Light).

Outstanding recognition was accorded a member of the 57th when Captain Emmett F. Knight, the 57th's Operations Officer, was selected as the Army Aviation Association of America's "Army Aviator of the Year" for 1962-1963. "Captain Knight's development of new and imaginative techniques for the employment of rotary-wing aircraft in the assault role contributed greatly to the success of helicopter operations in Vietnam."[4] The first Army aviator in Vietnam so honored, Captain Knight, in accepting the award, likened his role to that of a funnel, into which the individual talents and energies of all members of the unit were channeled into action. He further cited the maintenance and enlisted flight personnel and warrant officer pilots for their professionalism and dedication, critical element's

The 57th's first patch design was brought with the unit from Ft. Lewis, Washington and they continued to wear it for some time until a new unit patch was designed in 1962.[1]
Patch courtesy of Chris Miller.

The second unit patch was designed by CWO Allen B. Causseaux (nicknamed 'ABC') and was worn until June 1963.

98th Transportation Det. (CHFM) unit patch
Patch courtesy of Chris Miller

Novelty patch of the 98th Trans Det (CHFM)
Patch courtesy of Chris Miller

Above: **A Saigon street scene with the little French Renault taxis, known as "blue and whites", in which the first aviators of the 57th often rode to war.**
Photograph courtesy of Col. Emmett F. Knight, USA Ret.

Above Right: **A view of the 57th's tent city compound at Tan Son Nhut in early 1962.**
Photograph courtesy of Col. Emmett F. Knight, USA Ret.

Right: **The original 57th's Officer's Club, known as the "Shawnee Teepee", decorated with an ARVN Airborne Brigade parachute and 'Captured' Viet Cong flags. Of these Col. Knight writes, "The first time over the ground nothing but Viet Cong flags flew in the breeze and nobody was shooting . . . the second trip, no flags and some shooting . . . a third trip, Vietnamese flags everywhere and everybody was shooting."**
Photograph courtesy of Col. Emmett F. Knight, USA Ret.

to the 57th's success. Captain Knight, himself an accomplished writer, was also the subject of one of Major Cherne's articles for the U.S. Army Aviation Digest, November 1963, titled "Your Orders ... train 25,000 Vietnamese troops in helicopter assault tactics in 90 days."

While Captain Knight and Major Cherne may be singled out by name, the hand that these individuals had, along with their counterparts in the other H-21 units and Major Ivan Slavich of the UTT Helicopter Company, in devising, under combat conditions, effective airmobile tactics, methods and procedures must be recognized. These individuals, without the benefit of time or staff support, conceived, tested and developed the basic concepts of both troop lift, vertical envelopment and armed helicopter support that, with little modification, were employed throughout the war.

Being stationed in Saigon the crewmen of the 57th had to endure the good-natured ribbing, with possibly a little envy thrown in, of being called a 'Saigon Commando' by MAAG advisors and others out in the 'field.'[5]

FACILITIES

Upon arrival in Vietnam the unit was based at Tan Son Nhut airfield, and the unit's first quarters were in the Rex Hotel, downtown Saigon, which, at that time, was no more than a shell, without water, electricity, elevators, or kitchen! Field kitchens in tents on the roof of the hotel proved to be quite unique. Civilian clothes were the rule while downtown, but an alert required changing into flight suits and heading off to war in the little Renault taxis.[6]

As soon as the unit's gear caught up to them, a tent city sprang up at Tan Son Nhut, the flight line area was located immediately adjacent, on the south side, to the approach end of Runway 25. The 98th's aircraft maintenance was set up in the old French hangars, flanked to the south by the Operations building and the perforated steel planking (PSP) constructed ramp area to the north.[7]

By late 1962 the unit was well established, the tents having been replaced by semi-permanent housing known as "hooches"; wooden framed structures with cement flooring, masonry or wood slats and wire screen mesh walls, and corrugated roofs. The unit could even claim the first Armed Forces chapel to have been built in Vietnam, the "Chapel of Inner Strength" dedicated to the 57th's first fatality, SP5 Richard Leialoha K. Ellis. For entertainment a theater showed movies nightly, and two clubs were available, the enlisted mens' known as the "Playboy Lounge" and the officers' club called the "Shawnee Teepee", which had been upgraded from its

Captain Bill Rudd, at right in the upper left photograph, the 57th's Maintenance Officer, oversees the recovery of old '049' arguably the most famous H-21 to have served in Vietnam. Vietnamese engineers built a raft and used steel I beams to inch the aircraft to the road and safety. Towing the aircraft through Dong Tam, Captain Bill Rudd can just be seen leaning out from the tow jeep to check on his aircraft.

Photographs courtesy of Col. William C. Rudd, USA Ret.

original tent structure, and had a grand re-opening in a new facility on 9 March 1963.[8]

OPERATIONS DECEMBER 1961 - JUNE 1963

The 57th achieved many 'firsts' in Vietnam. Shared, with the 8th Transportation Company (Lt Hel), being first helicopter unit to be deployed, and, in a first for Army aviation, first fly-off from a ship, the USNS Core; first U.S. Army helicopters to fly in Vietnam; first joint heliborne assault; first to fly joint mission with a sister service (U.S. Marines); first to fly joint mission with the VNAF; first to train ARVN troops in heliborne operations; and, among many other firsts in Vietnam, first U.S. Army helicopter brought down by the Viet Cong.

From January to June of 1962 the 57th was engaged in intense training of three ARVN divisions in methods and techniques of airmobile assault. Eventually teams of two aircraft visited ARVN division airstrips to conduct this training, flying the aircraft to the troops' location rather than moving the troops to the aircraft. All the while the 57th still found the time to support a number of combat assault missions, and to show the Marines of HMM-362 the ropes when they arrived at Soc Trang in April 1962. Unit was tasked with support of III Corps and the ARVN 5th and 21st Infantry Divisons.

In spite of the 57th's Maintenance Officer's, Captain William C. Rudd, and the 98th Transportation Detachment's best efforts, the H-21 was aging and the 57th experienced a total of eight engine failures during the first six weeks in Vietnam. One incident was extraordinary enough to rate another first for the unit. H-21C s/n 56-2049, which was later to be the last H-21 flown in Vietnam, experienced total engine failure in flight and the crew attempted autorotation onto a road, but landed short in a rice paddy. Concerns about VC activity necessitated that the aircraft be removed from the rice paddy before dark. A group of Vietnamese engineers came to the crew's aid, jacked the aircraft up, built a wooden platform under it and rolled the aircraft onto three steel I-beam tracks and pushed it onto the road. The aircraft, 'Old 049', was then towed by jeep, under Captain Rudd's watchful eye, through the town of Dong Tam in the Delta, to the airfield. The 57th claims the first unit to hold the ground contact highway travel record for helicopters in Vietnam, 18 kilometers![9]

Captain Bill Rudd, who was ably assisted by CWO James Eakins, was also the unit's test pilot and flew his share of combat missions. Colonel Rudd writes "I was very busy trying to maintain a big fleet of aging cargo helicopters," but he also adds "Our recovery rate for helicopters downed by gunfire or mechanical problems was very high, given the scale of difficulty - about 88%."[10] All things considered, that was a remarkable achievement!

The unit suffered its first fatality, crew chief SP5 Richard Ellis, in

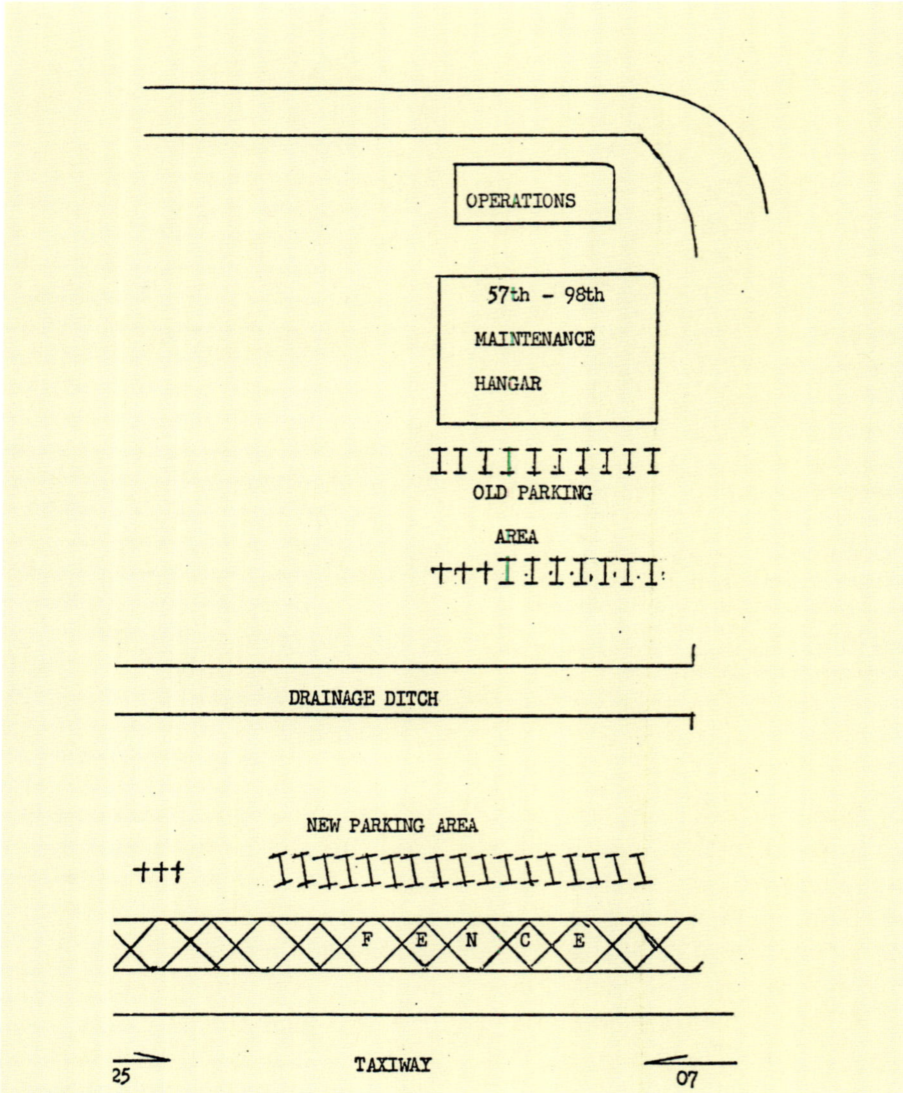

Figure 6. The 57th's flight line plan at Tan Son Nhut, early 1963, showing the Operations building, the old, and new, H-21 parking areas, and maintenance hangar. Adapted from the 57th Transportation Company (Lt Hel) history, 1 Jan to 23 June 1963, Appendix B.

Below: **ARVN troops training to board a 57th Trans. Co. H-21C s/n 56-2019 in early 1962. Aircraft carries authorized high visibility paint scheme and flight platoon markings. Flight platoon markings were not used in Vietnam by the 57th Trans. Co., and the markings on the aircraft were probably left over from Ft. Lewis.** US Army Photograph

an assault on Ap Bac on 5 October 1962. This first engagement at Ap Bac by the 57th Transportation Company (Lt Hel) did not receive much attention at that time, but, in addition to the one fatality, two aircraft were shot down. Colonel Knight, commenting on the selection of landing zones, and preplanned alternates, as it impacted the first "Ap Bac" writes "Neither the ARVN commanders nor their American advisors were fully competent to plan heliborne operations in those early days. (In spite of what LTC John Paul Vann thought.) ...what we did instead was simply land a little short, long, or left or right of the planned grid coordinates and the flight leader made that decision on final. ...very often this resulted in landing right on the warned and departing VC.... On the other hand, we often ran into trouble when the situation did not allow us to do that or when the flight leader was too doctrinaire or rigid to make that last minute adjustment. The only senior advisor who ever insisted on no deviation was Vann. He tried to control all decisions and that was a major cause of the Ap Bac before Ap Bac I know, and probably also had much to do with the 1963 imbroglio."[11]

The 57th engaged in support-

ing and assisting the ARVN in combat assaults for the remainder of 1962 and the first half of 1963. 2 January 1963 saw the 57th supporting Operation Burning Arrow, the assault on the Viet Cong's Central Office for South Vietnam in War Zone C. On 11 January 1963 the 57th participated in one of the longest heliborne assault lifts yet staged in Vietnam. Along with three aircraft of the 93rd Transportation Company (Lt Hel) the 57th airlifted ARVN 21st Division troops to an area near the Gulf of Siam, and later in the day returned to pick them up. Arriving back in Soc Trang late at night, a flight of six aircraft, after a brief stop, pushed on to Saigon. As the 57th's aircraft flew over the Mekong River at approximately 2200 hours, an aircraft suddenly fell out of formation, the rest of the flight watching in horror its red navigation lights weaving crazy patterns in the dark night sky, until it disappeared. The formation attempted to locate the downed aircraft, but after two hours of fruitless searching were ordered to cease their search. The next day rescue and recovery crews found the aircraft and seven dead crewmen and passengers, on an island in the Mekong.[12] It was a bitter start to the new year.

One aircraft recovery mission

More training, in this case with H-21C s/n 56-2049 which was to gain the distinction of being the last H-21C to be flown in the Republic of Vietnam on 27 June 1964, when LTC. Robert J. Dillard, the 57th's first CO in Vietnam, closed out the log book. US Army Photograph

Major Milton P. Cherne, left, Commanding Officer 57th Transportation Company (Lt Hel), confers with his Operations Officer, Captain Emmett F. Knight in a photograph taken early in 1962. The headgear, with its rather 'Australian' look, seems to have been popular with American advisors at the beginning of U.S. involvement. Major Cherne has a government issue Colt .45 in a decidedly non-government issue ammunition belt and holster, and both officers are armed with private purchase combat knives.

Photograph courtesy of Col. Emmett F. Knight, USA Ret.

Major Milton P. Cherne, CO of the 57th in his 'office'. This picture of a "tough looking, cigar-smoking warrior" to quote Col. Knight, "made such a hit in the Pentagon that it was displayed on the walls of the 'E' ring for several years." US Army Photograph

Aviators of the 57th take a break for chow in mid 1962, left to right, Lieutenant Rod Magee, CWO Ken Donley, CWO Don Youngblood, Captain Tom 'T.C.' West, CWO Frank Smith, Captain Emmett F. Knight, CWO Charley Larkin, CWO Winford "Giff" Gifford and Captain Dick Bastion. At least three of the pilots are wearing the second patch used by the 57th.

Photograph courtesy of Col. Emmett F. Knight, USA Ret.

in which the 98th Transportation Detachment (CHFM) built a 110 yard long track of perforated steel planking into a rice paddy to trundle two engines to two downed aircraft, and then changed both of their engines while under enemy fire and constantly having to return fire, resulted in the award of Bronze Star Medals with "V" device to all participants. On 11 April 1963 another engine failure resulted in an aircraft making a forced landing in a rice paddy, where the crew defended their aircraft from enemy attack until sufficient reinforcements had been brought in to secure the area. On 18 May 1963 the 57th participated in a very successful airmobile combat assault operation conducted against the infamous Viet Cong 514th Regional Battalion, which, caught by surprise, attempted to flee, and were cut down and resoundingly defeated, exacting a measure of revenge for Ap Bac.

Late in the afternoon on 11 June 1963, a day that became known to the 57th as "Black Tuesday" , a violent, tornado-strength wind swept through the unit's flight line at Tan Son Nhut, overturning and significantly damaging 12 H-21's and 4 TO-1D's. The 57th was out of commission; it was up to its successor unit, the 120th Aviation Company (Air Mobile Light), to put the pieces back together again.

To put in perspective the remarkable accomplishments of just one unit, it was calculated that the 57th Transportation Company (Lt Hel), and its successor unit the 120th Aviation Company (AML) flying the H-21, between December 1961 and December 1963 had "logged over 17,000 flight hours, over 35,000 cargo-ton-miles, over 1,000,000 passenger miles, and over 20,000 combat support sorties."[13]

An evocative photograph by SFC Al Chang, working for the Pacific Stars and Stripes, of the 57th's aircraft stacked in echelon right, on their way down to the Delta to pick up a load of troops. Visible serial numbers on the first two aircraft are 55-4140 and 56-2047. Date approximately April 1962.

Photograph courtesy of Col. Emmett F. Knight, USA Ret.

AIRCRAFT PAINTING AND MARKINGS

The 57th's aircraft arrived in Vietnam, from Ft. Lewis, Washington painted in overall Gloss Olive Drab (FS. 14087) with the Army's

Below: **H-21C s/n 56-2071 of the 57th exhibits the patchy look common after October 1962. It appears that this aircraft is half Gloss Olive Drab (FS. 14087) and the forward half Flat Olive Drab (FS. 34087) with a third, lighter color, painted over the ARMY marking on the fuselage. Date 8 February 1963.**

US Army Photograph

'Chicken Runner' a 57th Trans. Co. H-21C s/n 55-4143 flies high above the countryside on 8 February 1963. The painted out National Star Insignia is very visible. US Army Photograph

A rotor blade is replaced on H-21C s/n 55-4143 'Chicken Runner' on 21 February 1963. New rotor blades were frequently required due to the wooden rotor blades deteriorating in conditions of extremely high humidity. Some blades lasted less than ten hours. The 98th Transportation Detachment established a rotor blade repair shop and performed repairs on blades that would otherwise have been returned to the United States for repair. US Army Photograph

A 'gunship' of the 57th, H-21C s/n 56-2155, shows the installation of the XM-1 sub-system on 8 February 1963. The Shawnee gunship experiments were not particularly effective.
US Army Photograph

standard high visibility markings. Some aircraft carried flight platoon markings, on the nose and the vertical tail fairings, the two colors observed were yellow and white. Col. Emmett F. Knight recalls that even if flight platoon markings were carried in Ft. Lewis "they held little meaning for us and were pretty much gone in Vietnam. The aircraft were assigned to a specific platoon by TOE of course, but they were not

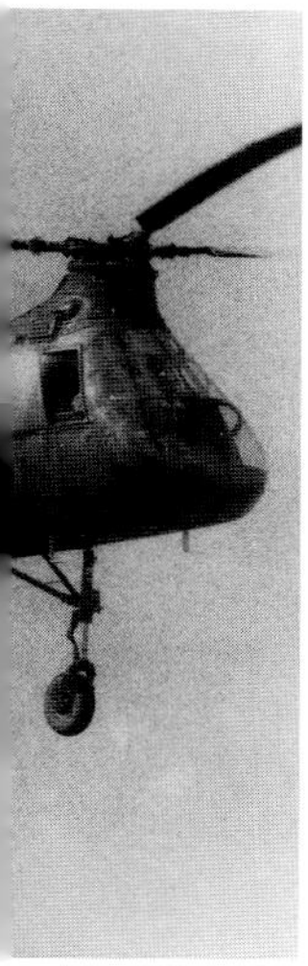

Superb view of four H-21's of the 57th Trans. Co. as they fly to the Plain of Reeds, from Tan Hiep airstrip on 13 March 1963. Nearest aircraft is H-21C, s/n 56-2055.
US Army Photograph

ARVN troops at the staging area of Ca Mau airstrip prior to being airlifted by the 57th Trans. Co. on 25 February 1963. In the background can be seen the YAC-1 Caribou, s/n 57-3080, being tested by the Ad-vanced Research Project Agency (ARPA). It was the only Army Caribou in Vietnam to carry the American flag on its tail.
US Army Photograph

Time magazine correspondent, Beverly Deepe, covering an airlift from Tan Hiep airstrip on 13 March 1963. CWO John G. Foley wears the blue grey flight suit.
US Army Photograph

employed that way in Vietnam. We used platoon and flight leaders and maintained continuity when possible, but we often assigned pilots to whatever tail numbers were available at the time and paired pilots and co-pilots from all company resources based primarily on the strength and weakness of each individual to get the best combination in the cockpit."[14] The colored trim remained on some aircraft until early 1963. A white, fuselage top conspicuity stripe was used by the 57th starting in approximately September 1962.

The 57th carried the 'last two' of the serial number on the forward rotor pylon in Gloss Yellow (FS. 13538) 8 inch high numbers. In one instance, where two aircraft shared the same 'last two', a third number was added to one of the aircraft, H-21C, s/n 56-2019, used '19' while H-21C, s/n 55-4219, used '219' on the forward rotor housing.[15]

The first three color camouflaged aircraft appeared in March 1963, and by May most of the 57th's aircraft were camouflaged.

In April 1963 the first unit designator, a white diamond with a red 57 in the center, appeared on the vertical tail of the 57th's aircraft.

In mid 1963 a few aircraft appeared with lower tail fairings painted in red, often with a thin white separation line. This may have signaled a re-introduction of flight platoon markings.

Many, if not most, of the unit's aircraft carried personal markings by early 1963. All 57th personal markings observed were painted just forward of the main, cargo door.

The unit's TO-1D's were painted Gloss Olive Drab (FS.14087) with the Army's standard high visibility markings. No unit markings were carried on the Bird Dogs.

Above: **A fine study of H-21C s/n 56-2048 of the 57th Trans. Co. with the 57th's white diamond unit marking visible on the vertical tail on 19 April 1963.** US Army Photograph

Right: **A three-color camouflaged H-21C s/n 56-2061 of the 57th Trans. Co. lifts off after off-loading ARVN troops on 14 March 1963. The three-color camouflage appeared on many of the 57th's H-21's in March 1963, and was in use through the year.** US Army Photograph

Above: **Vietnamese Marines, in tiger stripe fatigues, wait to board H-21's. The front two aircraft are from the 57th Trans. Co., H-21 s/n 56-2059 and s/n 56-2060 in the three-color camouflage. Aircraft s/n 56-2059, which wears the usual overpainted look, also carries, for the first time in April 1963, the 57th's unit designator, a white diamond with a red 57 in the center. This was the center design of the unit's second patch.** US Army Photograph

Top Right: **'Loose Goose' H-21C s/n 56-2047 of the 57th Trans. Co. roars in low over the ground, a gunner ready at the main door on 10 May 1963. On the original photograph the 57th's white diamond unit marking is just visible on the vertical tail.** US Army Photograph

Right: **On 11 June 1963, a day that was to become known as "Black Tuesday" to the 57th, a tornado-like wind destroyed 12 H-21's and 4 TO-1D's on the 57th's flight line. TO-1D s/n 57-2928 rests upside down at Tan Son Nhut.** US Army Photograph

One of the 57th's overturned H-21's at Tan Son Nhut after 'Black Tuesday' on 11 June 1963. Lieutenant Jon Stilman is in the foreground.

Photograph courtesy of David Prewitt

Above, Left and Far Left: **Three views of the 57th's aircraft carrying the three-color camouflage scheme in mid 1963. One is very artistically applied and the unique, light blue-green camouflage color is very apparent in the photograph above. The 57th's unit designator is also clearly shown. 'Chalk 14' has the personal marking 'Sexy' and an example of the hand brushed scheme. "Lucky" 'Chalk 13' has the personal marking 'Piasecki's Practical Joke' in red. Depending on the method of paint application, and thinning of the paint itself, the blue-green camouflage color is not as readily apparent in the two photographs to the left.**

Photographs courtesy of David Prewitt

Far Left: **A 57th Trans. Co. H-21C, s/n 56-2048, with toned down markings and yellow flight platoon markings on 1 February 1963.** US Army Photograph

Left: **The 57th's "Chapel of Inner Strength", at Tan Son Nhut during dedication ceremonies in March 1963.**

Photograph courtesy of Col. Robert L. Hoffman, USA Ret.

A composite profile of a 57th Trans. Co. (Lt Hel) H-21C, s/n 56-2110, in mid 1963, although this aircraft probably carried the rather untidy 'hand brushed' scheme. The last two digits of the aircraft's serial number, '10', were carried in yellow on the forward rotor pylon.

8th Transportation Company (Light Helicopter) (H-21)

140TH TRANSPORTATION DETACHMENT (CHFM)

8th Trans. Co. (Lt Hel) patch was brought with unit from Ft. Bragg, North Carolina.
Patch courtesy of Chris Miller

140th Trans. Det. (CHFM) patch 1961-1963
Patch courtesy of the US Army Transportation Museum

VIETNAM SERVICE

December 1961 to 23 June 1963 when the unit was inactivated. All personnel, assets and equipment of the 8th were transferred to the successor unit, the 117th Aviation Company (Air Mobile Light) along with the attached 140th Transportation Detachment.

COMMANDERS, NOTABLE UNIT MEMBERS, TRADITIONS

Major George D. Hardesty, Jr. was the commander of the unit during its remarkable ferry flight from Ft. Bragg, N.C. to Alameda Naval Air Station, California in November 1961. Major Hardesty was the first CO of the 8th in Vietnam, serving until mid year 1962 when Major Robert F. Corneil assumed command. Major Corneil was co-piloting the aircraft in which the unit's first combat deaths occurred, as outlined below.

Major Richard D. Kisling was the unit's third commander, serving until 20 April 1963, when he was promoted to become the Executive Officer of the 45th Transportation Battalion. Major Kisling "performed an outstanding job and leaves behind him a unit of excellent reputation." [1] The last Commander of the 8th was Major William S. Aiton, from the 52nd Aviation Battalion in Pleiku, who assumed command on Major Kisling's departure.

The 140th Transportation Detachment (CHFM) was critical to the unit's success, and all the pilots acknowledged the debt they owed these maintenance crews, who not only patched up their shot up birds, but went out to the field to bring home any aircraft that had been downed. One of the Detachment's early commanders was Captain William King, who was assisted by CWO 'Hound Dog' Wilson. [2]

CWO Henry "Hank" R. Beau was the unit's first combat casualty. CWO Beau was supporting an operation in the area of Cheo Reo when he was wounded by an enemy round that came up through the pilot's seat. CWO Beau was airlifted to Nha Trang, and then back to the United States where he ended up in Walter Reed Army Hospital and later had CWO Nysewander of the 93rd Transportation Company (Lt Hel) as a roommate. Nysewander had been wounded at Ap Bac in January 1963. CWO 'Hank' Beau authored the article, which told his story as well as Nysewander's, in the June 1963 'Argosy' magazine titled "America's Fighting Sky Cavalry." [3]

The 8th suffered its first combat deaths on 15 July 1962. A two ship flight had departed Kontum to reconnoiter a village which reportedly had been overrun by the enemy. One aircraft was piloted by CWO Joseph Goldberg, co-piloted by Major Corneil, the new CO of the 8th with SP5 Harold L. Guthrie crew chief and SP5 James E. Lane gunner and carried as an observer the senior American MAAG advisor in Kontum, Lt. Col. Anthony J. Tencza. On a low level run over the 'quiet' village the Viet Cong suddenly opened fire one round piercing the pilot's eye and, with no altitude to recover, the aircraft struck a tree which caused it to crash into a small horseshoe formed by the river near the village. Unable to land, the second aircraft returned to Kontum for assistance. Although hurt, with a spinal compression fracture, Major Corneil attempted to evacuate all crew and passengers from the aircraft and assisted SP5 Lane who was alive but wounded. Using the river for escape and evasion, SP5 Lane became separated from Major Corneil and did not survive. The 8th Transportation Company had, in the meantime, responded, and after some difficulty in securing ARVN troops, returned to the area where the aircraft piloted by Captain Joe D. Underwood was able to rescue the injured Major Corneil and recover the bodies of CWO Goldberg, SP5 Guthrie and Lieutenant Colonel Tencza, the MAAG advisor. SP5 Lane's body was not recovered until the following day, a mile or so down river.[4] The 8th Transportation Company (Lt Hel) honored and commemorated their own by naming the field at An Son, fifteen miles west of Qui Nhon, and two maintenance hangars in Qui Nhon after Goldberg, Lane and Guthrie.

FACILITIES

The 8th Transportation Company (Lt Hel), after the unit arrived in Saigon in December 1961, was initially based at Tan Son Nhut and billeted in downtown Saigon at the Tran Hung Dao Hotel. There they discovered that the lights worked occassionally, the elevator never, furniture and beds were missing and there was no kitchen or dining room. Like the 57th, the 8th Transportation Company erected a field kitchen on the hotel's roof!

Soon, on 10 January 1962, the 8th was relocated to Qui Nhon, where the Navy "Seabees" had constructed a base for the unit. A PSP flightline and runway for the helicopters, two hangars for maintenance and a tent city in downtown Qui Nhon became the unit's new quarters. Qui Nhon was a pleasant enough town on the South China Sea, and had its share of shops and bars dependant on the Americans.

Qui Nhon was to be the permanent home of the 8th, and for a long time its successor unit, the 117th Aviation Company (AML). Their downtown tent city was eventually relocated to a compound on the airbase, which was well laid out, with wide roads, and was to be described on 1 January 1963 as having "comfortable native-style bamboo hooches in what is probably the best-looking helicopter camp in the country." [5] The 8th did have two aircraft assigned, on an aircraft and crew rotating basis, in the Pleiku and Kontum area and two aircraft assigned to the Ban Me Thuot area. The aircraft always flew together, in case one went down, the second aircraft would provide as much assistance as possible.

Maintenance hangars of the 140th Transportation Detachment (CHFM) at Qui Nhon on 28 December 1962. Hangars are named to commemorate the first unit members killed in action, Lane and Guthrie. US Army Photograph

Above: **H-21C, s/n 52-8628, being serviced after sitting down on the beach near Tuy Hoa, Phu Yen Province on 14 September 1962. The distinctive look of the 8th's aircraft, after removal of the vertical stabilizers, is apparent. The new placement of the aircraft's serial number can be seen on the aircraft's tail cone.** US Army Photograph

Below: **Unusual, and rarely photographed, perspective of the top of the Shawnee's fuselage (aircraft s/n 52-8628) with a two color, white and yellow, conspicuity stripe. The H-21C, s/n 55-4216, in the background also carries the two color conspicuity stripe.** US Army Photograph

OPERATIONS

After the joint excercise and combat assaults with the 57th Transportation Company (Lt Hel) in the Saigon area, the 8th relocated to Qui Nhon, as mentioned, in early January 1962 to support ARVN forces in II Corps. One of the first major operations, and the largest that the unit had participated in to that date, was providing twelve aircraft in support of the ARVN 42nd Regiment in a two day operation which resulted in 40 Viet Cong dead on 30 March 1962.

With its two aircraft in Pleiku and two more in Ban Me Thuot, the 8th provided training and support to the ARVN troops in their area of operations. This not only included combat assaults, but also resupply of bases with ammunition and rations, and medical evacuation as needed. They also provided support to the U.S. Special Forces camps in the Central Highlands.

AIRCRAFT PAINTING AND MARKINGS

The 8th Transportation Company (Lt Hel) arrived in Vietnam with all H-21's painted in Gloss Olive Drab (FS. 14087) and the Army's standard high visibility markings.

Shortly after arriving in Qui Nhon to support II Corps, the 8th removed their aircraft's vertical stabilizers, to reduce weight for operations in the Central Highlands.

Since the H-21's serial number was carried on the vertical stabilizer, its removal by the 8th Transportation Company necessitated repositioning the aircraft's serial number and the U.S. ARMY lettering. The 'last three' of the serial number were then painted in white on the tail cone, the base color denoting, it is thought, the

Left: **PFC Kenneth R. Perrett poses with the .30 caliber machine gun at the main cargo door of a Shawnee with the 8th Trans. Co. (Lt Hel) on 28 December 1962. The aircraft carries the three color camouflage of light blue-green and tan over the gloss olive drab basecoat. Camouflage finish appears to have been sprayed-on as opposed to brushed on.**

US Army Photograph

Below: **Members of the 140th Transportation Detachment (CHFM) working on H-21C, s/n 55-4213, of the 8th Trans. Co. (Lt Hel) on 28 December 1962. It appears that the 8th were the first cargo helicopter unit to camouflage their aircraft in Vietnam. Note the fairings constructed to cover the tail stubs and the unit's unique placement of U.S. ARMY and the serial number.**

US Army Photograph

flight platoon, the whole thinly outlined in white. U.S. ARMY was marked vertically in yellow, to read from top to bottom, above the serial number on both sides of the rear fuselage.

A two color, white and yellow, conspicuity stripe was painted on the H-21's fuselage top by September 1962.

By December 1962 the unit had adopted a camouflage for their helicopters, believed to be the tan and light blue-green over the base gloss olive drab that was to distinguish the camouflage colors in this period in time.

Personal markings were carried forward of the main cargo door on the 8th's aircraft, one such personal marking noted was a camouflaged aircraft named "Swamp Fox 2", aircraft s/n 52-8648, the lettering at a slant to the fuselage.

93rd Transportation Company (Light Helicopter) (H-21)

UNIT NICKNAME 'SOC TRANG TIGERS' LATE 1962-1963
80TH TRANSPORTATION DETACHMENT (CHFM)

93rd Trans. Co. (Lt Hel) patch 1962-1963. "The 'mule', actually a burro, depicted on the patch had been the unit mascot at Fort Devens."[1]
Patch courtesy
CW4 Frank Baldwin, USA Ret.

Patch of the 80th Trans. Det. (CHFM) 1962-1963.
Patch courtesy of Chris Miller

VIETNAM SERVICE

January 1962 to 23 June 1963 when the unit was inactivated. All personnel, assets and equipment of the 93rd were transferred to the successor unit, the 121st Aviation Company (Air Mobile Light) along with the attached 80th Transportation Detachment.

COMMANDERS, NOTABLE UNIT MEMBERS, TRADITIONS

Major James E. Gray, Jr., who had commanded the unit at Fort Devens, Massachusetts, deployed the 93rd to Vietnam. His XO was Captain Don E. Finney and Captain Raymond DeLahanty was his Operations Officer. Major Gray, finding his unit stationed at the "end of the world", used his previous US Army experiences flying in Alaska to improvise, innovate and invent if need be, the tactics, procedures and means necessary to operate in mountainous I Corps. Major Gray was unlucky enough to lose his ship during Operation Lam Son II on 30 August 1962 in I Corps. Both he and his copilot on that mission, Lieutenant Lewis Stone, were wounded. Major Gray's downed helicopter's drive shaft was to become a permanent wall fixture at the Officer's Club in Soc Trang.[2] Major Gray rotated home to the United States soon after Operation Lam Son II and Captain Chuck Benedict became the acting CO until Major Paul R. Ewing assumed command in November 1962.

In late January 1963 General Earle G. Wheeler, the Army Chief of Staff, presented the first awards to the aircrews of the 93rd Transportation Company (Lt Hel) for heroism displayed at Ap Bac, pilots and gunners alike being cited for their actions.

The 80th Trans. Det.'s (CHFM) first Commander, 1st Lieutenant Kenneth C. Eaton, and the 93rd's Maintenance Officer, Captain Gary R. Bill, and then Captain Don Toth, had an unenviable task for it seemed that the elements, an aging helicopter, the critical supply situation and the Viet Cong all conspired to thwart their efforts. However in the early years in Vietnam it was Army aviation's good fortune that the adverse conditions encountered elicited extraordinary achievements from the men charged with keeping the Shawnees in the air. Captain Toth, as did his counterparts in the other H-21 units, exemplified that rare breed. He was well liked and respected by all who knew him, and figured prominently in veteran reporter Richard Tregaskis' book 'Vietnam Diary'. A good friend, then Captain Sam Conley, CO of the 611th Transportation Company (DS), recalls talking to Captain Toth before he caught a flight up to Saigon, reminding him of his commitment to visit Cap St. Jacques (Vung Tau) the former French beach resort area in Vietnam. Captain Toth answered that he'd get there soon, but that he'd promised Lieutenant Lew Stone to accompany him to Saigon.[3] Captain Toth was to die tragically the night of 11 January 1963 when the H-21, in which he was a passenger, crashed.

After the 93rd's aircraft had been toned down, one intrepid aviator, CWO Robert Kibler, determined to show 'the colors', carried an inscribed American flag in the cockpit of his H-21 on missions.

First company mascot was a nondescript Terrier named 'Tiger' who flew on numerous missions

An almost full grown 'Tuffy', the Bengal Tiger mascot, pet and pride, of the 93rd in mid 1963.
Photograph courtesy of Col. Kenneth D. Mertel, USA Ret.

with the 93rd out of Soc Trang in November 1962.[4] Shortly after this the 93rd acquired it's second mascot, 'Tuffy' a real Bengal Tiger. Tuffy was certainly the embodiment of the 93rd's nickname, the 'Soc Trang Tigers'. As a cub, the tiger had been acquired by a certain Captain R. A. Jones in Vientiane, and Tuffy promptly proceeded to chew up the cabin of his new owner's L-23. Captain Jones, with the MAAG Flight Detachment, Laos, took his new 'pet' home, whereupon his neighbors, obviously not being pet lovers, called the police. Jones then brought the young tiger cub to Vietnam and gave him to the 93rd.[5] Tuffy became an inevitable 'photo opportunity' stop in Soc Trang for all visiting dignitaries and VIP's, the braver of whom would actually pose with him as Tuffy was not declawed; fortunately Tuffy liked Americans! Tuffy eventually grew too large and had to be shipped, in mid 1963, to the Toledo Zoo, Toledo, Ohio by the 45th Transportation Battalion's Special Order #999, via Pan American Airways.[6]

FACILITIES

The 93rd Transportation Company (Lt Hel) had arrived at Da Nang air base in late January 1962, in the middle of the rainy season. They were assigned old French hangars on one side of the airfield, the town was once home to a French naval base and called Tourane, with perforated steel planking covered flight line and a small Operations building. Housing was in a military compound some distance away, where the unit quickly set about refurbishing assigned concrete barracks and redecorating at least one of their bathrooms with a large '93' set into the tiled floor.[7]

'Tuffy' was an eight month old cub when acquired by the 93rd Transportation Company (Lt Hel) as their mascot.

Right: **Captain Leon W. Curry, the Operations Officer of the 93rd Transportation Company (Lt Hel), poses with Tuffy in front of the 93rd's unit sign. Captain Curry was issued an Air Force flight suit, and, he stated, "wore Navy issue leather flight gloves", due to shortages in the supply of Army flying clothing.[8] Captain Curry was awarded the Distinguished Flying Cross for heroism at Ap Bac.**

Da Nang, with a population of a little over 100,000 was a good town for shopping, a small section of the town developed specializing in sales of jewelry, clothing and embroidered items to Americans, and recreation, like the 'Forget-Me-Not' Bar reputed to have the finest bar girls in Da Nang. For swimming and sunbathing, a favorite spot was 'China Beach', only eight miles away.

In September 1962 the 93rd switched locations with the Marine helicopter squadron at Soc Trang HMM -163, known as 'Rathburn's Ridge Runners' after their CO Lt. Col. Robert L. Rathburn. Soc Trang airfield had originally been built by the Japanese Army Air Force, and then was used by the French during the First Indochina War. The field consisted of a single runway with large parking aprons on either side, maintenance hangars, and Opera-

Above: **Captain Paul H. Brown, Executive Officer of the 93rd, keeps an eye, and a firm hold, on Tuffy as the cub explores a jeep.**

Bottom Right: **Tuffy being fed meat and kitchen scraps by, it is believed, PFC Michael D. Ferrante, the keeper of the cat. Tuffy obviously thrived on this diet as he has grown to the extent that others kept a safe distance when Tuffy was being fed, except one of the unit's dogs whose hunger seems to have overcome any caution.**

All photographs on this page courtesy of Col. Kenneth D. Mertel, USA Ret.

tions, Mess, a central shower facility and barracks buildings. The facilities were more spartan than up north, although for a change the water supply actually worked. The base was sometimes derisively called 'Bugville' by the Americans, due to the nightly swarms of black insects, green insects, bugs everywhere. The small town of Soc Trang was periodically off-limits to the Americans. Two landmarks in town were the Catholic orphanage, which was principally supported by the officers and men of the 93rd, and "the Graveyard" the area's most famous bordello, which acquired its name due to its location next to a cemetery.[9]

OPERATIONS FEBRUARY 1962 - JUNE 1963

The 93rd Transportation Company's (Lt Hel) H-21's flew off the USS Princeton (LPH-5), in bad weather conditions, from about ten miles out to sea when they deployed to Da Nang. The unit became operational on 1 February 1962. Arriving during the middle of the rainy season, the 93rd learned to make the most of breaks in the weather. The unit's mission was to support the ARVN in I Corps and the re-supply of outposts, both Vietnamese and U.S. Special Forces. A training program in heliborne operations for the ARVN was initiated almost immediately during the first 30 days of operation.

Aerial re-supply had an immediate, positive effect on the military situation in I Corps. Difficulties in getting support and supplies had created an increasingly unstable situation amongst the ARVN, which was alleviated by the 93rd. The stability that the unit brought even encouraged the ARVN to resume operations in the A Shau Valley, long an enemy stronghold, by mid year 1962. Logistical support of outposts meant delivering everything from pigs, ducks and chickens to howitzers, riding below the H-21 attached to slings, and even bulldozers flown in to a Special Forces outpost disassembled.

The unit gained valuable operational experience during the summer and still managed to host various American reporters, such as Dickey Chapelle, noted World War Two correspondent and visiting VIP's. General Hightower, on 27 June 1962, selected the 93rd as the most outstanding unit he had visited in the Republic of Vietnam. It was around this time that the unit also conducted a 'heliborne raid', lead by Major Gray, which lasted into the night; ARVN troops being landed on hillsides behind retreating VC in darkness such that the helicopter's landing lights had to be turned on to ensure a safe touch down. The 93rd believed that this was the first nighttime helicopter operation flown in Vietnam.

In early August 1962 the unit conducted the largest heliborne assault yet seen in I Corps, when two ARVN companies were airlifted in to raid the Viet Cong 5th Regional Headquarters.

However it was the 30 August 1962 combat assault, Operation Lam Son II, that remained vividly in the memories of the 93rd. A joint operation was to be carried out with the VNAF 1st Helicopter Squadron (H-34); the target, a Viet Cong Battalion, had been warned of the raid and was covering the

ARVN troops board H-21's of the 93rd, aircraft 'S' and 'Y', s/n 51-15908, are visible, with blue flight platoon markings, Da Nang, May 1962. US Army Photograph

Off-loading troops in I Corps, a 93rd H-21C, aircraft 'I', s/n 51-15902, in standard high visibility markings, and flight platoon colored trim, has just set

landing zone with large caliber weapons. Skyraiders of the VNAF provided close air support and worked the area over before the lifts touched down. Bad weather then delayed the airlift, and when the weather cleared, the enemy was ready for the helicopters. Eight out of twelve H-21's were hit on touch down, and two of them brought down. The 93rd's CO's, Major Gray, aircraft was shot down as mentioned above, and he was rescued, along with his crew, by two pilots of the UH-1A Flight Section who witnessed the Major's aircraft 'explode', killing three ARVN on board. The H-21 was hit by what may have been an enemy .57 recoilless rifle. Four other aircraft that were hit in the landing zone were damaged enough to be considered disabled.

With Lam Son II still fresh in their minds, an advance team left Da Nang 12 September on a classified mission to relocate the 93rd to Soc Trang in the Delta. This was accomplished on 19 September 1962, when HMM-163 and the 93rd Transportation Company (Lt Hel) exchanged locations. The move was not an easy one, with little support except from other services, and as the 93rd settled into Soc Trang, all of its aircraft were grounded for maintenance. It took one or two weeks, into late October, for the unit to become combat ready again and begin operations in support of the 7th ARVN Infantry Division in IV Corps. The unit was to be sorely tested on 2 January 1963, at Ap Bac, in which the 93rd lost a crewman, SP4 Donald L. Braman, and had six aircraft downed and the UTT lost a crew chief, and one UH-1B, during the battle.[10] A major turning point for Army aviation in Vietnam, this battle's significance became apparent when the rules of engagement were changed, and Army aviators entered combat in earnest. A second blow to the company occurred a few days later on 11 January 1963 when three of its own lost their lives, as passengers, in the crash of a H-21 from the 57th Transportation Company, (Lt Hel).

The 93rd Transportation Company (Lt Hel), during the first six months of 1963, continued their high level support of the ARVN in airmobile operations, flying supplies to the Sea Swallows in Binh Hung, Father Hoa's village in the Delta, and supplying aircraft for joint operations with the other H-21 units. The 121st Aviation Company (Air Mobile Light) proudly assumed the name of the famed "Soc Trang Tigers" in late June 1963 as the successor unit to the 93rd.

AIRCRAFT PAINTING AND MARKINGS

The 93rd Trans. Co.'s (Lt Hel) aircraft were painted in Gloss Olive Drab (FS. 14087) with the standard high visibility markings when they arrived in Da Nang in January 1962. Major Gray allowed flight platoon colors of blue and red, with white trim, to be carried by the aircraft on the nose triangle formed by the canopy framing, as well as on the tail fairings, top and bottom. Photographs of the red flight platoon color are rare.[11] Later, possibly as part of 'toning down' the markings, the 93rd dropped the color red, substituting white, as a flight platoon identifier on the nose and tail fairings. These

down outside of enemy 'anti-helicopter' stakes.
US Army Photograph

H-21C, aircraft 'B', s/n believed to be 52-8664, lands during a training exercise in Da Nang, May 1962.
US Army Photograph

Right: **Pilots of the 93rd pose on 19 December 1962 in front of a Shawnee with toned down markings, the painted out National Star Insignia being readily apparent. Kneeling, left to right, CWO's Thomas W. Barnes, Jr., Keith R. Borck, John Fuller and standing CWO's Henry Singer, Sidney B. Shows and Bobby E. Raulston. Manning the door gun is PFC James Latimer. CWO Borck was to be awarded the Distinguished Service Cross for his actions, and CWO's Barnes, Raulston and Singer were to be awarded DFC's for heroism at Ap Bac. CWO Shows is still wearing on his cap, and maybe his flight suit, the patch of his former unit, the 13th Transportation Company (Lt Hel), in Korea. The personal armament of Swedish K's and Thompson sub-machine guns are of note.**
US Army Photograph

Left: **A rather patchy looking 93rd H-21C, s/n 52-8624, aircraft 'N' on a take-off run from Tan Hiep airstrip on 13 March 1963. The aircraft appears to be painted in overall Gloss Olive Drab (FS. 14087) with high visibility markings painted out in flat olive drab. The reduced size individual letter 'N' is painted in yellow on the tail cone. The individual aircraft letter positioned on the forward rotor pylon remained the same size as before but was repainted in yellow. A red conspicuity stripe is just visible on top of the fuselage and a personal marking of a small boy is carried in front of the main cargo door.**
US Army Photograph

latter two platoon colors, blue and white, carried over into the 121st Aviation Company (AML).

The 93rd Transportation Company (Lt Hel) was the only H-21 unit to use letters for aircraft identification in Vietnam. A 24 inch high letter in white was carried on the tail and a 12 inch high letter on the forward rotor pylon. Around the time of the change of location from Da Nang to Soc Trang, the high visibility markings were toned down and the aircraft's large tail identification letter was reduced to approximately 10 inches in height and repositioned on the aircraft's tail cone, while the forward rotor pylon letter remained the same size. The color of the aircraft identification letter was changed from white to yellow, however, both colors, white and yellow, were observed in use, side by side, well into 1963.

ARVN troops wait to board a 93rd H-21C, aircraft 'G', s/n 52-8655 (the number 8 has a bullet hole patch on it which makes it appear to be a 3), on 22 March 1963. The aircraft is very clean looking, painted in overall gloss olive drab. US Army Photograph

Many of the 93rd's aircraft carried a red 'conspicuity' stripe on top of the fuselage from mid 1962 on, instead of a white stripe.[12] The stripe's color quickly took on the appearance of a pale, faded red due to the weathering action of sun, rain and the wear and tear of operations.

CW4 Frank Baldwin, USA Ret., former member of the 93rd, recalled that the 93rd may also have camouflaged their aircraft prior to inactivation on 23 June 1963, but there is no widespread photographic evidence. The successor unit, the 121st Aviation Company (AML) did fly some camouflaged H-21's until early June 1964, with the tiger's head unit insignia painted on the vertical stabilizers.

The 93rd was the only H-21 unit observed in which personal markings were carried behind the pilot and co-pilot's windows.

The commander of a USAF T-28 'Farm Gate' Detachment, Major Doyle, hosted a barbecue party at Soc Trang. Left to right, Captain Curry, Captain Brown, Captain Anderson, Major Doyle and unknown.
Photograph courtesy of LTC. Leon W. Curry, USA Ret.

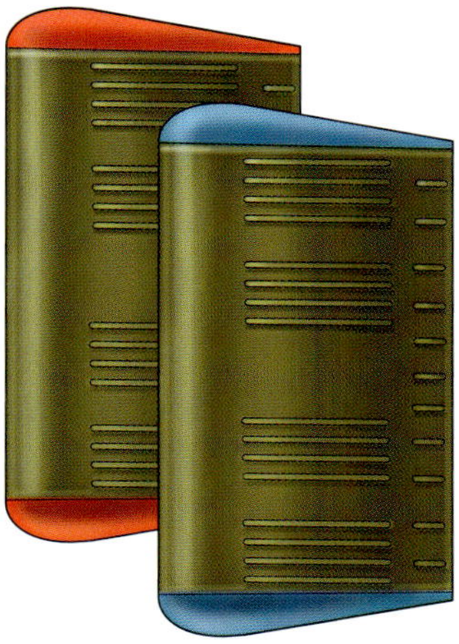

Left: **Early 93rd Flight Platoon colors of red and blue as trim on tail fairings, January - late 1962.**

Right: **93rd Flight Platoon colors of blue and white as trim on tail fairings, late 1962- June 1963.**

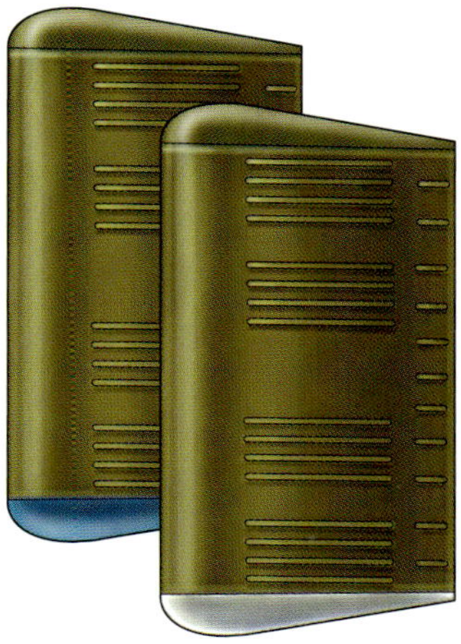

A particularly pleasing in-flight view of 93rd Transportation Company (Lt Hel) H-21's in early 1963. Note the red conspicuity stripes on the aircraft in the foreground.

Photograph courtesy of CW4 Frank Baldwin, USA Ret.

(Note: In the 93rd Trans. Co. (Lt Hel) Flight Platoon colors were also carried on the nose triangle formed by the cockpit canopy framing. Neither the U.S. Army and aircraft serial numbers in yellow, nor the early 24" high white letter designator on the tail fins are depicted, for the sake of clarity.)

Illustrated by Steve Davis

Left: **In front of the assembled Company on 24 April 1963, Major General Lawrence J. Lincoln, Deputy Commanding General USARYIS, awards CWO Thomas W. Barnes, Jr. the Distinguished Flying Cross for heroism at Ap Bac, 2 January 1963. Both CWO Barnes and CWO William T. Lamb, also awarded the DFC that day, wear dark blue "Sea Swallows" 'beercan' insignia on their personalized utility caps (CWO Barnes' swallow insignia reflects the sun and so appears to be silver). CWO Lamb is wearing the 93rd patch on his right fatigue pocket.** US Army Photograph

Below: **Father Hoa's "Sea Swallows" patch, 1963. Referring to Binh Hung in the Delta the Vietnamese wording translates as "The Secret Place of the Sea Swallows."**

Patch courtesy of CW4 Frank Baldwin, USA Ret.

33rd Transportation Company (Light Helicopter) (H-21)

UNIT NICKNAME 'BALM-DE-BALM' 1962-1963

573RD TRANSPORTATION DETACHMENT (CHFM)

Patch of the 33rd Trans. Co. (Lt Hel)
Patch courtesy of Chris Miller

Patch of the 573rd Trans. Det. (CHFM)[1]
Patch courtesy of the US Army Transportation Museum

VIETNAM SERVICE

September 1962 to 23 June 1963 when the unit was inactivated. All personnel, assets and equipment of the 33rd were transferred to the successor unit, the 118th Aviation Company (Air Mobile Light) along with the attached 573rd Transportation Detachment.

COMMANDERS, NOTABLE UNIT MEMBERS, TRADITIONS

The 33rd's Commanding Officer, Major Joseph E. Henderson had assumed command of the unit on 1 November 1960 and was the only commander of the 33rd Trans. Co. to have served in Vietnam. The 33rd compiled an outstanding record under Major Henderson, no personnel or aircraft lost to the enemy. Several aircraft were shot down, but they were recovered successfully.

Major Henderson and eleven other officers, warrants and enlisted men, were awarded the Air Medal, for twenty five completed missions, on 11 April 1963 by Lieutenant Colonel Hoffman, Commander of the 45th Transportation Battalion. Captain Ben E. Luck was the 33rd's Operations Officer.

The 33rd went about its business without fanfare and built a solid reputation as a unit that could be counted on to achieve its mission. The 33rd was an outstanding unit, one of the best of its time.

The unit nickname was derived from a popular Vietnamese beer known as '33', which carried a 33 within a circle on the label. Vietnamese for 33 is "Ba Muoi Ba" or, as pronounced by most Americans "Balm-De-Balm", which was how the beer became known, as well as the unit.[2]

The 573rd Transportation Detachment (CHFM) maintenance area at Bien Hoa in June 1963. The 573rd performed all their work in the open, without benefit of hangars. H-21C, s/n 56-2134, is being servied in the background. Note the M-38A1 jeep's wheel cover with the unit insignia and '33rd Maintenance Officer' inscribed on it.
Photograph courtesy of Col. Robert L. Hoffman, USA Ret.

FACILITIES

The 33rd Transportation Company (Lt Hel) was based at Bien Hoa, and was assigned an area of the air base which was, at that time, little more than a dirt strip. The unit set up tents in a compound on the air base, the aircraft's flightline being a grass field. The 573rd Transportation Detachment (CHFM) had to make do in the open, exposed to the elements, there being no hangar facilities available for aircraft maintenance. Needless to say their facilities were described as very dusty.[3]

OPERATIONS

When the 33rd Transportation Company (Lt Hel), stationed at Fort Ord, California was alerted for the second time in 1962, the first alert having occurred on 6 March 1962 for deployment on 18 April, it was not a false alarm. The main body of the Company departed California, by C-121, on 10 September 1962 and flew to Saigon. The aircraft were shipped to Saigon on the USNS Croatan arriving on 18 September 1962. The Military Assistance Command, Vietnam had estimated that the unit would take 8-10 weeks before they could become operational, the 33rd accomplished it in early October, in a little over two weeks.

Then Major Henderson recalls "We had no assigned area - we ranged from Da Nang in the North to the 93rd's area in the South."[4]

AIRCRAFT PAINTING AND MARKINGS

The 33rd had given up five aircraft to the 8th Transportation Company (Lt Hel) in November 1961, and to other units deploying until all 20 H-21's that they had been so proud of at Fort Ord were gone. When alerted for their Vietnam deployment they had to get 20 replacement aircraft from all over the United States. The 33rd's replacement aircraft arrived in Vietnam in Gloss Olive Drab (FS.14087) with the Army's standard high visibility markings. As soon as they landed at their home station, Bien Hoa Air Base, the high visibility markings were painted over in gloss olive drab, and the unit's aircraft operated in toned down markings, generally without flight platoon colors or unit markings, although a 'left-over' marking, from a previous aircraft user unit, or two may have existed.

According to Colonel Henderson "I do not remember using any special markings on our aircraft. In fact, we probably never got them all marked up..."[5]

By early 1963 personal markings may have been painted on some of the unit's aircraft.

H-21C, s/n 52-8642, of the 33rd Transportation Company (Lt Hel) being readied on the flight deck of the USNS Croatan on 18 September 1962. The aircraft is painted in overall Gloss Olive Drab (FS. 14087) with the Army's standard high visibility markings. The cargo door would be removed as soon as the aircraft reached Bien Hoa Air Base. US Army Photograph

View of the flightline of the 33rd Transportation Company (Lt Hel) at Bien Hoa Air base in September 1962, H-21C, s/n 52-8644, in the foreground. Aircraft have had their markings 'toned down' and no flight platoon or unit markings were carried. US Army Photograph

Above: **A neatly painted 33rd Trans. Co. (Lt Hel) unit insignia mounted on a fence in the company area.**
Photograph courtesy of LTC. Joseph Henderson, USA Ret.

Right: **Flight Operations building and sign of the 33rd at Bien Hoa.**
Photograph courtesy of LTC. Joseph Henderson, USA Ret.

Major Joseph Henderson, Commander of the 33rd Trans. Co. (Lt Hel), sits at the end of the table at the Maintenance and Commanders conference in Vung Tau, May 1963. From left to right, Major Mathews, unknown, Major Beauchamp, Major Henderson, unknown, Brigadier General Stilwell, LTC Klingenhagen and LTC Robertson.
Photograph courtesy of Col. Sam Conley, Jr., JSA Ret.

81st Trans. Co. (Lt Hel) patch came with the unit from Hawaii, 1962-1963

Insignia of the 545th Trans. Det. (CHFM) 1962-mid 1963 has not been identified or documented

81st Transportation Company (Light Helicopter) (H-21)

545TH TRANSPORTATION DETACHMENT (CHFM)

VIETNAM SERVICE

September 1962 to 23 June 1963 when the unit was inactivated. All personnel, assets and equipment of the 81st were transferred to the successor unit, the 119th Aviation Company (Air Mobile Light) along with the attached 545th Transportation Detachment.

COMMANDERS, NOTABLE UNIT MEMBERS, TRADITIONS

Soon after the 81st Transportation Company (Lt Hel) arrived in Pleiku, in September, Captain Paul F. Anderson was the acting commander of the unit and a true innovator. It was under his guidance that the 81st experimented with H-21 gunships. In the unit's first couple of months in Pleiku the unit had no aircraft hit and no casualties. Major George Washington Aldrich, Jr. next assumed command of the unit and was present on the mission on which CWO Holloway was killed.

FACILITIES

Located in the plateau country of the Central Highlands of Vietnam, the old airstrip at Pleiku was dominated by a large, flat-topped hill that contained the regional military headquarters structure that was known, when the 81st was initially based in Pleiku, as 'the Citadel.'[1] The flight line was covered with perforated steel planking (PSP) and the unit's tent city compound was at the edge of the airstrip, at first named 'The Country Club'. The 545th had four or so rudimentary hangars from which to work. So primitive were the accomodations in fact that Pleiku was often referred to as "Devil's Island" by Army aviators because it was General Stilwell's favorite place to 'exile' those who did not live up to his expectations.[2] The arrival of the Headquarters of the 52nd Aviation Battalion in March 1963 certainly helped to improve living conditions.

OPERATIONS

The 81st took over the support of a number of remote U.S. Special Forces locations in the Central Highlands of II Corps, such as Plei Yit and Plei Mrong from the 8th Transportation Company (Lt Hel). One of the unit's early missions involved rescuing an ARVN Battalion which was attempting a river crossing when caught in the middle of the river by a flash flood. The 81st responded late at night, with a VNAF flare ship to provide lighting. Ground fog put an end to the night operation, but at first light the unit flew back to the river and, in two lifts, pulled out some 60 to 70 troops who had been clinging to logs or were still in the water. Many had died in the flood. After this action Captain Anderson, acting CO, wryly discussed the high 'pucker factor' associated with navigating at night and being so close to the Laotian border.[3]

On 16 October 1962 the 81st participated in their first, large scale operation when 400 ARVN's were airlifted into a VC training camp area where some 69 enemy were killed, and an enemy battalion commander was captured along with 30 tons of rice.

The 81st had their first fatality on 22 December 1962 when they conducted a joint mission with the 8th Transportation Company (Lt Hel). Each unit was tasked to furnish fifteen aircraft, but due to maintenance problems, the 81st arrived with only fourteen H-21's, so there were 29 altogether on the combat assault. Picking up ARVN troops in Tuy Hoa, the aircraft flew at 2,000 feet towards the target area, which had six different, prearranged landing zones. While a Bird Dog marked one of the LZ's with red smoke, a ground fire warning was radioed to the helicopters now approaching at contour level. As soon as the H-21's flared for landing, the Viet Cong opened fire, hitting six aircraft out of nine in the one LZ. Crewing the eighth ship of the 81st's flight, were CWO Charles E. Holloway, CWO Daniel Gessang, SP6 James Mooney, crew chief and SP4 Arthur M. Whitmore, gunner. Their ship was hit hard just after off-loading their troops, rounds smashing into the cockpit hitting CWO Holloway in the forehead, just missing CWO Gessang, snapping SP4 Whitmore's, the gunner, safety harness and three rounds penetrating the fuel tank. As CWO Gessang desperately raced the H-21 to Tuy Hoa, the crew chief attempted to bandage CWO Holloway's wound and plug the fuel tank with his gloves. As soon as they landed CWO Holloway was placed on a litter but died before he could be taken to the 8th Field Hospital in Nha Trang.[4] The unit was shaken by his death, he left a wife and five children behind, and immediately started raising funds for his family. The Army named Camp Holloway, shortly to be home to the 52nd Aviation Battalion, in Pleiku after CWO Holloway as a memorial. General Stilwell dedicated the Camp on 4 July 1963.[5]

The landing zones had not been prepped by either artillery or airstrikes and many in the unit thought that was an oversight which contributed to their ships getting shot up.

AIRCRAFT PAINTING AND MARKINGS

The 81st deployed to Vietnam, from supporting the 25th Infantry Division in Hawaii, with their aircraft painted in overall Gloss Olive Drab (FS. 14087) with the Army's standard high visibility markings. They were the only H-21 unit to arrive in Vietnam with unit insignia painted on their aircraft. In addition to the unit's 'grasshopper' insignia they also carried flight platoon colors as trim on the nose triangle formed by the cockpit framing, and the vertical tail fairings. Flight platoon colors were white and, it is believed, red. The 'last three' of the serial number were often carried within the flight platoon nose marking.

The aircraft were quickly 'toned down' losing their high visibility markings, but retaining the flight platoon colors and the 'grasshopper' unit insignia on the vertical tail.

The 81st is believed to have used a white stripe conspicuity marking on their aircraft.

Personal markings were painted just forward of the cargo door.

Two more Shawnee's are being readied for flight. H-21C, s/n 56-2054, carries white platoon markings on the nose and as trim on the tail. The light gray upper surfaces and flat black undersides of the aircraft's rotor blades can be seen in the foreground.

US Army Photograph

H-21C, s/n 56-2020, of the 81st Trans. Co. (Lt Hel) left foreground carries the 'last three of the serial number within a white flight platoon nose marking. Both the 81st and the 33rd Trans. Co.'s (Lt Hel) arrived in the Port of Saigon on the USNS Croatan in the background on 18 September 1962.

US Army Photograph

Far Left: **81st Trans. Co. (Lt Hel) H-21C's, s/n's 55-4145 and 51-15896, lift-off on the first leg of their journey to their final destination, Pleiku.**

US Army Photograph

Left: **General Earle G. Wheeler addresses the officers and men of the 81st Transportation Company (Lt Hel) on 20 January 1963. In the background H-21C, s/n 56-2017, has had the high visibility markings removed although not the flight platoon markings, the unit's 'grasshopper' insignia and carries a personal marking forward of the cargo door.**

US Army Photograph

H-21C, s/n 56-2020, of the 81st Trans. Co. (Lt Hel) is painted in overall Gloss Olive Drab (FS. 14087). White flight platoon markings are carried on the nose, with the 'last three' of the serial number in black, and on the tail fairings. The unit's insignia is carried on the tail under the serial number.

Utility Tactical Transport Helicopter Company

UTT patch from Okinawa 1962
Patch courtesy of Chris Miller

Second design UTT patch from Vietnam, 1963-1964
Patch courtesy of Chris Miller

571st Transportation Detachment (Acft Maint) 1962-1964
Patch courtesy of the US Army Transportation Museum

UNIT NICKNAME "UTT" 1962-1964

HEADQUARTERS PLATOON CALL SIGN "SABRE" 1963-1964

1ST PLATOON CALL SIGN "PLAYBOY" ONE THROUGH SIX 1963-1964

2ND PLATOON CALL SIGN "RAIDERS" TWO-ONE THROUGH TWO-SIX 1963-1964

3RD PLATOON CALL SIGN "DRAGONS" THREE-ONE THROUGH THREE-SIX 1963-1964

571ST TRANSPORTATION DETACHMENT (AIRCRAFT MAINTENANCE)

VIETNAM SERVICE

October 1962 to 14 August 1964 when unit was inactivated. All personnel, assets and equipment of the UTT were transferred to the 68th Aviation Company (Armed Helicopter) which was activated on 15 August 1964.

COMMANDERS, NOTABLE UNIT MEMBERS, TRADITIONS

The Utility Tactical Transport Helicopter Company, known as the UTT, had been activated in Okinawa on 25 July 1961.[1] Prior to being sent to Vietnam, the first armed helicopter unit in the Army, spent months working up in the Philippines and then in Thailand. On 9 October 1962 the UTT deployed to Vietnam under the command of Major Robert E. Runkle. Major Runkle, well liked by his men, introduced the UH-1A to combat, but favored a conservative approach in utilizing his aircraft. His successor, Major Ivan Slavich assumed command on 21 November 1962 and instituted a more aggressive use of his 'fighter helicopters.' Major Slavich is credited, by many of the early members of the UTT, as being the driving force behind the unit's success. He was a superb pilot, a former Marine and paratrooper with combat experience in Korea, and had a quite, decisive manner which translated into his being an outstanding combat leader. His was not a foolhardy type of leadership, in fact he was happiest when the H-21's he escorted did not even take fire, but he inspired in his unit 'the spirit of

Perched on a raised rice paddy levee like so many dragonflies, a flight platoon of the Utility Tactical Transport Helicopter Company composed of two UH-1A's and three UH-1B's, serial numbers visible 62-1886 and 62-1880, on stand down between missions in the late 1962 to early 1963 time frame. Photograph by Captain Roger Baker via Bob Chenoweth

The color photographs on the following two pages were taken by Captain Roger Baker who served with the UTT in 1962 and into 1963. All photographs via Bob Chenoweth

Right: **The UTT compound, beautifully maintained, at Tan Son Nhut. Typical hooches line both sides of the gravel walkway.**

Below: **View inside the UTT Flight Operations buiding at Tan Son Nhut.**

Below: **UTT Maintenance, Operations and Armament building.**

Above: **Aircraft maintenance area, with tents and hangars, as seen from the flightline utilized by the UTT at Tan Son Nhut.**

Right: **The UTT's 'Auger Inn' Officer's Club with a unique signpost, a UH-1A tail fin, s/n 59-1687. Later unit histories only admit to one UH-1A being lost by the UTT in its first year, and it is this aircraft that has led many historians to conclude that the only aircraft lost by the UTT at Ap Bac on 2 January 1963, was a UH-1A. The story behind this specific aircraft's loss and how its tail fin came to adorn the Officer's Club is still somewhat of a mystery.**

December 1962 and the UTT's first Christmas in Vietnam. An 'Open House' was held for the Vietnamese.

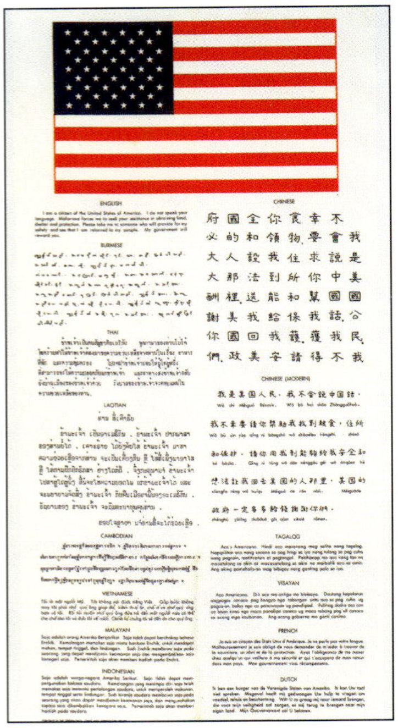

USAF Aeronautical Chart and Information Center (ACIC) April 1961 XM-61-1 'Blood Chit'. The English text is translated into thirteen (two separate Chinese) languages and reads: I am a citizen of the United States of America. I do not speak your language. Misfortune forces me to seek your assistance in obtaining food, shelter and protection. Please take me to someone who will provide for my safety and see that I am returned to my people. My government will reward you.

Collection of Chris Miller

During the 'Open House' a village chief is outfitted with a flak jacket as the UTT does it's part to 'win hearts and minds.' A rare, early 'blood chit' used in Vietnam can be seen on the back of the flak jacket.

A 'Christmas' pig is swung on board a UTT UH-1B. US Correspondent Beverly Deepe photographs the activity for the audience back home, most of whom would be still hard pressed to locate Vietnam on a world map in December 1962.

Right: The UTT provided a reminder of home and the Holidays with Christmas lights strung on the unit's sign and a small Christmas tree.

Far Right: First Christmas in Vietnam and Santa has forgone his reindeer and sleigh for a more technologically up-to-date transport, a UH-1B, as envisioned by the UTT.

attack' that separates true fighter units from the rest. Major Slavich's Executive Officer was Captain Gene Burress.

The UTT received the Army Aviation Association of America's 1963 "Outstanding Army Aviation Unit" Trophy for "making outstanding contributions to or innovations in the employment of Army aviation on and beyond the normal mission assigned."[2]

On 22 June 1963 Major Robert Reuter assumed command of the UTT, and led the unit through the summer and fall of 1963 as the pace of engagement and the size of enemy units encountered increased. The UH-1A's had been withdrawn by mid year and the unit was now fully equipped with UH-1B's. On 16 December 1963 Captain Calvin R. Bean became the new Commanding Officer, and soon after the First Flight Platoon rejoined the unit at Tan Son Nhut from Da Nang. Airmobile operations continued in support of the ARVN, and as more Army aviation units arrived in country, the number of participating units increased, setting new records for heliborne assaults.

Early work on the armament systems, eventually employed so successfully by the UTT, were the responsibility of a few key individuals. CWO Clarence J. Carter, in January 1962 on Okinawa, developed a skid mounted machine gun system for the UH-1A. CWO Cleatus Heck built the first UH-1A rocket launcher pods of eight rockets each. A 2.75 folding fin aerial rocket (FFAR) system was devised by Major Luther Lollar and CWO Heck. CWO Clemuel H. Womack also designed a low drag 2.75 FFAR launcher and initiated the work which would result in the M-16 system. The UTT also pioneered the suspension of the M-60 machine gun from the cabin roof with an elastic bungee cord.[3]

The UTT was inactivated on 14 August 1964. Because the UTT was established for testing purposes, and was a TDA unit, the unit could not and did not accrue campaign participation or decorations. The Chief of Military History determined that under the provisions of paragraph 10-4d, AR 870-5, lineage and honors are determined for TOE units only which are flag, color or guidon bearing. In fact the UTT is not considered by the Army to have a common lineage with any other Army aviation unit in Vietnam! The Office of the Chief of Military History went on to state that "the 68th, 197th and 334th Aviation Companies are separate and unrelated organizations with different histories."[4] Colonel William W. Fraker, USA Ret. (formerly Playboy 16, Raider 26 and Sabre 3) wrote in the July 1974 issue of U.S. Army Aviation Digest "the response from.....the Office of the Chief of Military History, are good examples of the difference between the real world and officialdom. In the real world the UTT, 68th, 197th and 334th most assuredly have a common lineage and by no stretch of the imagination can it be said that the 68th, 197th and 334th are separate and unrelated organizations with different histories."[5] A tradition was born with the UTT and its spirit did indeed live on through all the Sabres, Playboys, Raiders and Dragons of the UTT and later years.

FACILITIES

The UTT had their own tent city compound at Tan Son Nhut airbase, which by November 1962 they had started to upgrade. Later 'hooches' were constructed and the compound improved in both appearance, and more importantly, comfort. An effort was made the first Christmas in Vietnam to decorate in the spirit of the season, and the unit arranged an 'open house' to be held for the Vietnamese. The flight line was shared, on either side, by the USAF and the VNAF, although the 571st Transportation Detachment (Aircraft Maintenance) did have a hangar. The unit's headquarters was based at Tan Son Nhut throughout its service in Vietnam, however the First Platoon was sent to Da Nang in February 1963 and didn't rejoin the unit until December 1963.

OPERATIONS

The UTT arrived in Vietnam with fifteen UH-1A helicopters, organized into 3 flight platoons of five helicopters each. Each flight platoon had an assigned color, red for the 1st Platoon, white for the 2nd Platoon and blue for the 3rd Platoon. Col. Emmett Knight later noted "the UTT came to stay with us.... red leader this is blue leader.... blue leader this is red...etc., etc., etc., escort chatter was so high for awhile that the 57th flight leader could hardly get a word in edgewise."[6]

Due to the rules of engagement and the armed escort nature of the UTT's role, an ARVN observer was required to be carried in each helicopter. In theory the ARVN representative was supposed to identify a target as Viet Cong and approve firing back if fired upon. Due to the fact that many ARVN could be Viet Cong themselves, or sympathizers, they were unarmed as a safety precaution. Many, however, did not speak English and often communication was nonexistent.

At first the mere presence of the UH-1A's of the UTT seemed to be enough to discourage any enemy response to the H-21's that the Hueys were escorting. However the enemy studied helicopters tactics and began to train their troops in ways to defeat the 'Dipper' as they called the UH-1A/B.[7]

The UTT's first fatality occurred on 3 November 1962 when CWO Richard 'Pappy' H. Wright's, the unit's veteran pilot, door gunner was mortally wounded. An escort mission was underway, and the second lift was being escorted into the LZ when the enemy opened up with automatic weapons. On CWO Wright's second pass, SGT Jonnie G. Lee, the gunner, called out that he'd been hit, a round had just missed the groin protector he was sitting on and severed his leg artery. SGT Lee died within minutes of CWO Wright landing to get him medical attention.[8]

The Army Concept Team in Vietnam (ACTIV) was actively involved in evaluating the armed helicopter concept from 16 October 1962 to 15 March 1963. While the term "escort" was not well defined "actual experience determined that the escort role broke down into an enroute phase, that was generally flown at a relatively safe altitude, the approach phase, where the heliborne force usually descended to nap-of-earth heights several kilometers away from the landing zone, and the landing zone phase. It was in the landing zone phase that the armed helicopters proved most valuable."[9] ACTIV was able to show that the incidents of fire, in the landing zone phase, against UTT escorted H-21's actually dropped, while hits, or Viet Cong effectiveness, against unescorted aircraft doubled! The conclusion was that suppressive fire delivered by armed escorts was highly effective, a conclusion seconded by the H-21 units.

The unit's second fatality occurred at Ap Bac on 2 January 1963, when crew chief Sergeant William L. Deal was killed in the crash of the UTT UH-1B, s/n 62-1882, downed by enemy gunners. For his part 1st Lieutenant James E. Stone, the UTT flight platoon leader, was awarded the Distinguished Flying Cross. His citation reads, in part , "Lt. Stone's UH-1B helicopter crashed after being hit by Viet Cong fire as he was attempting to land to rescue the crew of the two other downed helicopters. He was knocked unconcious in the crash. Lt. Stone's selfless act of attempting rescue in the face of blistering Viet Cong fire was an act of extreme heroism."[10] Ap Bac was the heaviest fire fight that the UTT had yet been engaged in, and led to the following conclusion regarding the Huey's of the UTT that "You can't use these things as you would a tank. When you start trying to assault a strongpoint at low altitude, and almost hovering, you're asking for hits. And these hits may kill you."[11]

The unit split up in February 1963, the First Flight Platoon relocating to support the Marine helicopter unit in I Corps, while the Second and Third Flight platoons continued to fly escort for the 57th, the 33rd and the 93rd Trans. Co.'s as well as providing reconnaissance and convoy protection and close support for ARVN ground troops. The UTT set up a system to locate one of the platoons with a supported unit, the 93rd Transportation Company (Lt Hel) in Soc Trang for example, then rotating a second platoon to relieve them a week later. The unit worked with the ARVN to expand the number of 'Eagle Flights' conducted, as well as, during April through July 1963, taking the newly arrived UH-1B unit's, such as the 114th Aviation Company (Air Mobile Light), 'gun' platoons under their wing to assist them install weapons systems and train their personnel.

When the 1st Platoon rejoined the unit in December 1963, the UTT was structured with four platoons of five aircraft each; a Headquarters Platoon 'Sabres', 1st Platoon 'Playboys', 2nd Platoon 'Raiders' and 3rd Platoon 'Dragons'. The unit continued to rotated platoons, for a week at a time, through supported unit's bases, such as Vinh Long airfield in the Delta.

AIRCRAFT PAINTING AND MARKINGS

The first UH-1A's that the UTT received were delivered in Gloss Olive Drab (FS. 14087) and carried the Army's standard high visibility markings. The UTT had devised a Flight Platoon marking system for their 15 UH-1A's while on Okinawa. This system used a color code to designate each flight platoon, red for the 1st, white for the 2nd and blue for the 3rd. The aircraft were marked with an appropriately colored circle on the nose and a tail fin band, upon which the individual aircraft's flight platoon number was painted in a contrasting color, i.e. white number on red colored base for the First Platoon.

Within a couple of weeks all the unit's UH-1A's had their colorful markings over-painted in Gloss Olive Drab, leaving only the serial number and ARMY marking in Gloss Yellow (FS. 13538) on the tail fin, and the roof and belly National Star Insignia.

UH-1A's often carried personal markings in the form of names on the small removable panels that had previously carried the crew chief's name. "Marj" and "Ole Hotbox" are two that have been observed.[12]

In November 1962 the UTT received its first eleven UH-1B's which were delivered in Gloss Olive Drab and the usual high visibility

A superb shot of UH-1A, s/n 59-1673, aircraft #2 of the 1st Flight Platoon of the UTT in Thailand on 20 July 1962. The aircraft is painted overall Gloss Olive Drab (FS. 14087) with standard high visibility markings. The 1st Flight Platoon color, red, is carried as a band on the tail fin and as a circle on the nose. A white 2 is centered on both the band and the circle. The first UTT unit insignia is carried below the crew chief's name on the door post; altogether a very colorful aircraft.

US Army Photograph

An aircraft nicknamed 'Marj' showing the typical placement of personal markings on the UH-1A. US Army Photograph

UH-1A. s/n 59-1695, photographed on 21 February 1963 in support of Operation Guy Phoung 9. US Army Photograph

Above: **UH-1A's on the flightline at Tan Son Nhut**

Left: **Captain Roger Baker stands by UH-1A, s/n believed to be 59-1673, First Platoon of the UTT in overall gloss olive drab.**

All photographs on this page by Captain Roger Baker via Bob Chenoweth

markings. The UH-1B's did not have their markings toned down, and operated alongside the over-painted UH-1A's until mid 1963 when the UH-1A's were retired.

The Army Concept Team in Vietnam (ACTIV) evaluated camouflage on at least three of the UTT's UH-1B's.[13] Most photographs show a three-color camouflage pattern, the colors appeared to be an olive green or tan and black over the base olive drab. One aircraft may have carried a simplified, two-color camouflage scheme of olive green over the base olive drab in early 1964. There does not appear to have been any guidelines for the application of the camouflage for each aircraft had a different pattern. The most consistent camouflage pattern seen is a three-color scheme of slanted, alternating bands of color along the length of the aircraft.

By late 1963, all UTT aircraft had acquired a unit marking of yellow crossed cavalry sabers on the pilot's and co-pilot's doors.

Armor plate kits were introduced of Aluminum-Doron panels, to protect the pilot's and co-pilot's seats and as add-on panels for the crew doors and the cargo doors.[14] The cargo door armor consisted of two separate panels, one attached to the door itself, almost half the height of the cargo door window, and a lower, unattached section of armor plate. Personal markings in the form of girl's names began to appear, painted on the cargo door window armor, around late 1963 and into 1964. The armored panel was often painted in a bright background color, such as red and sometimes was elaborately done, with a white trim, as on "Kathryn" UH-1B, s/n 62-2065, and sometimes painted simply as on "Pandora" UH-1B, s/n 62-1949.

The first UH-1B's being delivered to the UTT by USAF C-133 transports, 29 November 1962.

UH-1B s/n 62-1885, on the UTT flight line at Tan Son Nhut, March 1963.

All photographs on this page by Captain Roger Baker via Bob Chenoweth

Huey's of the UTT in low level approach to their flightline at Tan Son Nhut in this dramatic view.

An aerial view of the UTT's flightline at Tan Son Nhut

UH-1B, s/n 62-1884, is serviced on 21 February 1963, by SP6 Gordon M. Johnson and SP4 Almer D. Kelson at Tan Son Nhut.
US Army Photograph

UH-1B, s/n 62-1878, with add-on window armor, on the UTT flight line in late 1963.
Photograph via Jim Sprinkle

PFC Symkowski latches the TR (Top Right) gun into position on the XM-6 weapon system on 12 February 1964. He wears the second design UTT unit patch on his left fatigue pocket. US Army Photograph

Major Ivan Slavich, Commanding Officer, Utility Tactical Transport Helicopter Company in early 1963.
US Army Photograph

Above: **UH-1B, s/n 62-1949, of the UTT on the unit's flightline on 12 February 1964. Aircraft carries the personal marking 'Pandora' in white on, it is believed, a red background. At least two camouflaged UH-1B's, under test by ACTIV, are visible. Aircraft serial numbers visible to the right of 'Pandora' are 62-1951 and 62-4953.**

US Army Photograph

Right: **UH-1B, s/n 62-4593, of the UTT on 12 February 1964 with the unit's yellow crossed sabers marking. The six designates the commanding officer's aircraft, which also carries an extra FM antenna.**

US Army Photograph

Above: **UH-1B, s/n 62-1877, of the UTT in an experimental three color camouflage scheme under test by ACTIV on the flightline in March 1964. The three colors were tan and black over the base-coat olive drab. The UTT unit marking of crossed sabers is carried on the crew doors. Aircraft is armed with the XM-14/LAU-2 system. Note the name 'Big Bertha' painted on the XM-3 system on the UH-1B in the background.**

US Army Photograph

Above Left: **The three-color camouflage scheme of olive green and black over the base-coat of olive drab is shown to advantage on the UTT's UH-1B, s/n believed to be 62-1885, being refueled from a 1st Aviation Company (CV-2) Caribou at Phi Truong Buu Son in June 1963 on their way up to Da Nang. Major Slavich is standing by the crew door.**

Photograph courtesy of Col. Robert L. Hoffman, USA Ret.

Above: **Photographed during a mission in June 1963, UH-1B, s/n 62-1885, of the UTT painted in the three-color experimental camouflage scheme of olive green and black over olive drab.**

Photograph courtesy of Col. Robert L. Hoffman, USA Ret.

Above and Profile Below: **UH-1B, s/n 62-2065, in overall Gloss Olive Drab (FS. 14087) with standard high visibility markings, February 1964. Aircraft is armed with the XM-6E3 sub-system with a MA-2 rocket tube pack. The UTT's yellow crossed sabers are carried on the crew doors and a personal marking 'Kathryn' is painted in white on a red background, white cheat line, on the cargo window add-on armor. A second personal marking 'Little-Rudy' is painted in yellow on a MA-2 rocket tube.**

USAF Photograph

UH-1B, s/n 62-1877, one of the UTT aircraft to be evaluated by ACTIV in experimental camouflage. This particular scheme utilized black and tan applied in bands over the original olive drab basecoat. The unit's crossed sabers marking and the ARMY and last five of the serial number are painted in yellow on the tail fin. The aircraft is armed with the XM-14 .50 caliber system and a LAU-2 rocket pod.

23rd SWAD patch 1962-1964
Patch courtesy of John Cross

Grumman Corporation designed patch – US manufacture
Patch courtesy of John Cross

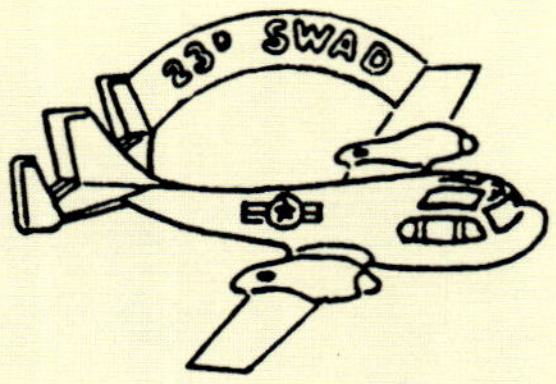

Small brass 'beercan' insignia of the 23rd SWAD

23rd Special Warfare Aviation Detachment

UNIT NICKNAME "THE WARRIORS" - 1963 & 1964

REPAIR SHOP NAMED "HAWKS LAIR" 1963

VIETNAM SERVICE

The 23rd Special Warfare Aviation Detachment (SWAD) arrived in Vietnam in October 1962. Unit served with distinction until inactivated and personnel, assets and equipment were combined with those of the 4th ASTA Platoon to form the 73rd Aviation Company (Aerial Surveillance) on 26 December 1964.

COMMANDERS, NOTABLE UNIT MEMBERS, TRADITIONS

Major William J. Morris was the first commander of the 23rd SWAD. Major Morris not only had to contend with the initial, very limited 'rules of engagement' for his aircraft, but his unit's very success generated more calls for support than his six aircraft could handle. As a high profile unit he also had to host the numerous VIP's who wanted to view his unit's operation. By the end of November 1962 the unit had been split into flight teams of two aircraft each, and repositioned where best able to support ARVN operations in II Corps.

The second unit commander was Major Liebl who oversaw the unit's Headquarters relocation from Nha Trang to Vung Tau in mid 1963. At an August 1964 Change of Command ceremony Major David B. King, II assumed command of the unit. In December 1964 Major King became the first commander of the 73rd Aviation Company (Aerial Surveillance) (OV-1).

FACILITIES

The 23rd SWAD was initially stationed at Nha Trang which was already home to the 339th Transportation Company (DS) and the 18th Aviation Company (U-1A). Nha Trang was another old French seaside resort town, with beautiful beaches, good restaurants and a tent city which provided adequate quarters. The Officer's club, established by the 18th Aviation Company (U-1) in a converted French Villa, was already well known as the 'Cockpit Club'.

After November 1962, one flight section was relocated to Qui Nhon to support the ARVN 9th Division, and soon another was repositioned to Pleiku, in the Central Highlands. The accommodations at Pleiku were primitive compared to Nha Trang or Qui Nhon.

Eventually, in mid 1963, the unit Headquarters was relocated to Vung Tau, although the unit still maintained flight sections in Soc Trang, Quang Ngai, Nha Trang, Qui Nhon etc.

OPERATIONS

The 23rd SWAD, a controversial unit from the beginning, was initially formed as a prototype armed surveillance unit over the objections of the USAF, which felt the Mohawk violated the intraservice close air support role agreements.

The 23rd SWAD was under the operational control of MACV and had been sent to Nha Trang on 16 October 1962 to support the ARVN 9th Infantry Division and the Railway Security Agency. The unit consisted of a Headquarters, three flight sections of two armed Grumman JOV-1C Mohawks, and a Photo Processing Section.[1] The JOV-1C was an OV-1C which had its infrared imaging gear removed and was equipped with wing hard points for weapons attachment.[2] On 23 November 1962 a flight section was moved to Qui Nhon.

Initially the Mohawks could only be armed with .50 caliber weapons, and could only fire if fired upon. As the rules of engagement were relaxed in the Spring of 1963, 2.75 inch folding fin rockets were added to the Mohawks' weapons stores.

A second flight section was positioned in Pleiku in early 1963, and one of their aircraft mysteriously disappeared on a routine mission in February 1963. The mystery of this disappearance was partially solved when the wreckage was spotted, in a remote mountainous area southeast of Pleiku, by a Bird Dog of the 73rd Aviation Company (Airplane Surveillance) (Light) in late December 1963.

As a ground search party, led by Lieutenant Colonel William C. Siebert, Commanding Officer of the 52nd Aviation Battalion, cut through the dense jungle to reach the crash site, helicopters of the 119th Aviation Company (AML) and VNAF T-28's provided top cover. The Mohawk had crashed and burned after cutting through the treetops, killing both crewmembers, but the cause of the crash was never determined.[3]

ACTIV was involved in this first test of the armed surveillance concept. Missions involved both visual and photographic reconnaissance for the supported units, and the Mohawk crews became intimately familiar with the terrain and population, able to spot enemy structures and insurgents in the open. "One of the unique advantages of the Mohawk in reconnaissance was

The 'Tent City' quarters of the 23rd SWAD in Nha Trang in November 1962. The small sign proclaims 'Best Tent of the Week'.
US Army Photograph

The 23rd SWAD mess building, left, and the dining hall in Nha Trang, November 1962.
US Army Photograph

Below: **JOV-1C, s/n unknown, aircraft #2 of the 23rd SWAD is towed to the maintenance hangar on 21 February 1963 in Nha Trang.**
US Army Photograph

Upper Right: **Aircraft #2 is guided into the 'Hawks Lair' maintenance shop by unit personnel. The shop sign is just visible behind the aircraft's left engine nacelle.**
US Army Photograph

Lower Right: **The 23rd SWAD's 'Hawks Lair' shop sign on 21 February 1963. Lieutenant John A. Kern, of the 23rd SWAD, wears a rabbit's foot for good luck.**
US Army Photograph

SP4 William H. Elliot loads 2.75 inch rockets into a SWU-7/A37 rocket pod on 21 February 1963. The .50 caliber weapon pod appears to have been 'borrowed, from T-28's.
US Army Photograph

SP4 Elliot places the switching units into place. His section and the unit designation are embroidered on his cap.
US Army Photograph

Below: **General James F. Collins, Commander in Chief US Army Pacific, visiting Nha Trang on 12 March 1963. The elaborate 23rd SWAD sign is noteworthy. General Collins is talking to Major Henry Q. Davis, Jr., Commanding Officer of the 18th Aviation Company (U-1A); the Otter insignia of the 18th is just visible on his right fatigue pocket. Brigadier General Stilwell is to General Collins' left.**
US Army Photograph

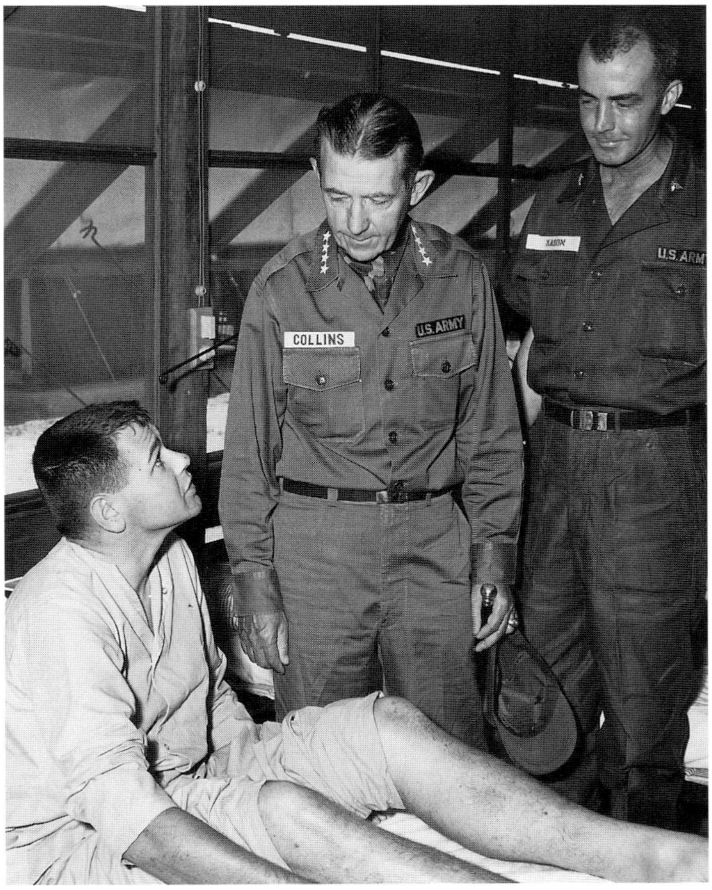

Left: **Lieutenant Edward Cribb, one of the crewmembers that ejected successfully from the 23rd SWAD Mohawk on 9 March 1963, is visited in the 8th Field Hospital, Nha Trang, by General Collins on 12 March 1963.**

US Army Photograph

Middle: **While the circumstances behind the cause of this crash are not known, this is a rare and unique photograph of a downed Mohawk which has 'bellied in', partially losing its right wingtip. While most of the airframe remains intact, very little remains of the cockpit. Believed to show the 23rd SWAD's second Mohawk loss on 9 March 1963. The crew appears to have ejected successfully.**

Photograph courtesy of John Cross

Below: **Line up of 23rd SWAD replacement aircraft in Nha Trang in mid to late 1963. Note the aircraft #12 on the nearest Mohawk.**

Photograph courtesy of John Cross

its speed to noise relationship which allowed the aircraft to get within observation distance of people on the ground without alerting them to its presence."[4] Incidents of Viet Cong attacks against the railway system decreased by fifty percent after regular surveillance was instituted by the 23rd SWAD.

On 9 March 1963 the unit lost another Mohawk, the crew ejecting, and only suffered relatively minor injuries. A third Mohawk, operating out of Soc Trang, also crashed as a result of a mid-air collision, around late March 1963, the crew being killed.[5] More aircraft were needed, to replace the three lost, and fill the increased tasking. American advisors, such as Colonel Hal D. McCown, Senior advisor in II Corps, were requesting a relaxation of the rules of engagement for the Mohawks. In his opinion, Mohawks armed with rockets, napalm and small fragmentation bombs would seriously hinder the Viet Cong's ability to conduct daylight operations, and he had been supported by only two aircraft![6]

"From 16 October 1962 to 15 March 1963 the 23rd Special Warfare Aviation Detachment flew more than 2,000 hours in the performance of 785 combat support missions. It had delivered defensive fire 27 times and had lost two aircraft."[7]

AIRCRAFT PAINTING AND MARKINGS

The Mohawks of the 23rd SWAD were painted in overall Gloss Olive Drab (FS. 14087) with the Army's standard high visibility markings, as explained in the text of Section IV. Each of the first six JOV-1C aircraft which deployed with the 23rd SWAD were marked with a yellow (FS. 13538) number, from one to six, on the tail in 24 inch high figures and a smaller, 12

Above: **At his Change of Command ceremony Major Liebl delivers his farewell address to the 23rd SWAD in Vung Tau, August 1964 as Lieutenant Colonel Robert J. Dillard, commander of the Aviation Support Battalion (Provisional), listens. The Mohawk in the background carries 14 red hearts on a yellow rectangle.**

Photograph courtesy of Col. Robert J. Dillard, USA Ret.

Far Left: **The 23rd SWAD's second commander, Major Liebl, wearing the unit patch in this informal portrait.**

Photograph courtesy of Col. Robert J. Dillard, USA Ret.

Left: **Captain Mitchell, Maintenance Officer of the 23rd SWAD, photographed in front of the 611th Trans. Co. (DS) fixed wing repair hangar at Vung Tau in February 1963.**

Photograph courtesy of Col. Sam Conley, Jr., USA Ret.

A 23rd SWAD group photograph in late 1964. Major David B. King, II, third unit commander, is standing to the right of the photograph. In the background aircraft #17, s/n unknown. Photograph courtesy of Col. Robert J. Dillard, USA Ret.

23rd SWAD Mohawk, believed to be an OV-1A, s/n unknown, with 'Mohawk Indian' emblem on the nose. This aircraft was photographed in Nha Trang mid to late 1963 as part of a line up of replacement aircraft received by the 23rd SWAD.

Photograph courtesy of John Cross

Close up of the 'Mohawk Indian' emblem. This may not have been a 23rd SWAD marking, but may have been left over from a previous assignment.

Photograph courtesy of John Cross

inch high number, was painted on the nose for aircraft identification purposes. Once the unit started to receive additional OV-1A aircraft, the numbers appear to have been continued in sequence.

The early JOV-1C aircraft did not carry any unit marking other than the aircraft identification numbers on the nose and tail. Towards mid to late 1963, at least one of the unit's OV-1A's was observed with a round 'Mohawk Indian' insignia on the left side of the nose.

There was a 23rd SWAD aircraft marking practice of painting small, red hearts on a yellow, rectangular field on the left side of the nose presumably to record enemy hits that the aircraft had taken. Two aircraft, flying out of Soc Trang, were observed, in late 1963 or early 1964, to have carried markings on their aircraft of small red hearts on large, yellow, rectangular fields.[8] Another OV-1 Mohawk, in the background of the August 1964 Change of Command ceremony, was also photographed with this unique unit marking.

Above and Profile Below: **JOV-1C, s/n 62-5857, 23rd SWAD aircraft #5 sits on the apron at Soc Trang in late 1963. The aircraft is painted in overall Gloss Olive Drab (FS. 14087) and carries the Army's standard high visibility markings. Additionally the aircraft carries a yellow rectangle with 14 red hearts painted on the left side of the nose, a unique unit marking. Aircraft is armed with an XM 157 rocket pod with an aerodynamic nose cone.**

Photograph courtesy of John Cross

611th Transportation Company (Direct Support)

First 611th Trans. Co. (DS) patch design, 1963-1965
Patch courtesy of the US Army Transportation Museum

A fine study of Captain Conley at the controls of his unit's U-1A Otter, in early 1963.
Photograph courtesy of Col. Sam Conley, Jr. USA Ret.

UNIT NICKNAME "BLACKHATS" 1962

UNIT NICKNAME "AXES" 1963

UNIT CALL SIGN "MOONSHINE" 1962-1963

VIETNAM SERVICE

The 611th arrived in Vietnam in October 1962; later became part of the 34th General Support Group and served with distinction until the end of Army aviation involvement in the Vietnam War in 1973.

COMMANDERS, NOTABLE UNIT MEMBERS, TRADITIONS

The 611th Transportation Company (DS) deployed to Vietnam under the command of Captain Ed Davis, who, within a week, was to return to the United States for medical reasons. The 611th was then briefly commanded by Captain Jim Blackmore before Captain Samuel G. Conley, Jr. assumed command in early December 1962.[1] Within a couple of months the 611th, under Captain Sam Conley, had developed into one of the most highly decorated Army aviation units, for its size, in Vietnam with an outstanding reputation and record. There is no question that the 611th often had a high profile mission; every aircraft downed by the enemy or that crashed, always attracted attention, most of it unwanted, from a higher command, from a sometimes critical press or from enemy snipers. In spite of having to work under the pressure of so much scrutiny, with the right combination of determination, mental toughness, hard work and leadership Captain Sam Conley managed to instill a unique esprit into his unit, so much so that it was to be said of the 611th "The word "gung ho" is seldom applied to a Transportation Field Maintenance Company (DS); however, this may be another "first" for Army aviation for such a company does exist here in Vietnam." [2]

The Recovery Platoon Leader Captain Milton C. Shattuck, Jr. and the 611th's Maintenance Officer CWO John F. Leonard, Jr. both earned acclaim for themselves and the unit. CWO Leonard was awarded his first Distinguished Flying Cross for his role in evacuating downed aircraft in the Ap Bac area during the operation and while under fire. He acquired his second Distinguished Flying Cross (First Oak Leaf Cluster) a month later "for heroism while participating in aerial flight: Warrant Officer Leonard distinguished himself by heroic action on 6 February 1963 while serving as the co-pilot of a CH-21C helicopter in the Republic of Vietnam. Warrant Officer Leonard volunteered to fly ammunition and supplies to a Vietnamese outpost besieged by Viet Cong forces and evacuate the critically wounded in his unarmed CH-21C helicopter. Fully realizing the hazards involved, he flew at maximum gross load to the outpost where the surrounded Vietnamese were greatly outnumbered and were receiving heavy machine gun and mortar fire. Although continuously threatened by hostile fire, Warrant Officer Leonard and his crew unloaded the ammunition and supplies, and loaded twenty-six casualties aboard the helicopter. Through the skill of Warrant Officer Leonard, a successful take-off was accomplished although the aircraft was critically loaded and under hostile fire......"[3] Captain Shattuck was the pilot of that aircraft and he too received his First Oak Leaf Cluster to the Distinguished Flying Cross.

While pilots often grabbed the headlines, and the glory, in the 611th everyone, from clerks to cooks, wanted to be involved and would plead with the Recovery Team Chief, Sergeant First Class Ivon G. Coleman, to be allowed to go on the next aircraft recovery mission. Sergeant Coleman, himself awarded the Distinguished Flying Cross, took his recovery teams anywhere and everywhere to fix up, cut up or make ready to fly out downed helicopters and fixed wing aircraft, and successfully towed by truck, floated by riverboat, or hauled out, by ox cart if necessary, damaged aircraft and engines so that they could be put back in the air. It appears that from the highest to the lowest rank, in the 611th, enthusiasm, professionalism, courage, a 'can do' attitude and aviation skill were the standard, not the exception and the unit proved each day it was the "Combat Maintenance Company".[4]

Captain Conley outfitted everyone in the unit with black hats, hence the nickname "Blackhats", with the lettering '611th Trans Co (DS)' embroidered in gold thread on the front in an arc, and the wearer's job title added under the unit designation. It was a real morale booster.

The second unit nickname "Axes" was acquired at Ap Bac in early January 1963. After the battle, but still under sporadic enemy sniper fire, the 611th was engaged in recovering the 93rd Transportation Company's downed H-21's. Most combat damaged aircraft were either temporarily repaired or had engines or components replaced before being flown out. One downed 93rd H-21C, s/n 51-15908, was made ready for flight on 4 January 1963, and as then Captain Conley recalls, he was about to help Captain Carl Strever, Maintenance Officer of the 45th Transportation Battalion, fly the aircraft out, when, at the last minute, Captain Strever determined to fly solo. In attempting lift-off, Captain Strever struck a rice paddy levee and rolled the aircraft, thrashing the blades on the ground. The aircraft was beyond repair on site. Captain Conley, who had never abandoned an aircraft to the enemy or destroyed one to prevent it falling into enemy hands, made the decision to cut the aircraft in two, at the fuselage field splice. Using axes the 611th recovery crew chopped the H-21 in half, and both sections were airlifted out by the Recovery Platoon's H-21, s/n 56-2134. Thereafter the 611th Transportation Company (DS) was called the "Axes", a nickname the unit adopted as a badge of honor, even commemorating their new name by having the unit's artist paint a picture of an eagle, clasping an ax in each talon, on the Recovery Platoon's H-21![5]

Upon Captain Conley's return to the United States in September 1963, Captain Charles E. Nicholls assumed command of the unit, building on the unit's fine reputation. The 611th, under Captain Nicholls' command helped prepare, in November and early December 1963, the Caribou's of the 1st Aviation Company for redeployment back to the United States.

FACILITIES

The 611th Transportation Company (DS) was the first Army aviation unit to be based in what was then still known as Cap St. Jacques, the old, French name for Vung Tau. The unit's three assigned aircraft, an Otter U-1A, Beaver U-6A and the Recovery Platoon's H-21 staged out of Vung Tau in support of the Army aviation units in the Delta. Utilizing the familiar old French hangars for aircraft maintenance and repair work, and lining up their machine and tool shop shelters outside the hangars, the 611th was in business. Like most units in Vietnam, the 611th utilized Vietnamese workers, known as local nationals, which were assigned various, non-technical, duties such as building H-21 cradles. The CO recalled that their methods were often primitive, and nothing was completed at speed, however they did get the job done.[6]

The Orderly Room, mess hall, which boasted a huge unit emblem imbedded in the floor, clubs and enlisted housing were all on the air base. Lack of adequate quarters required the officers and warrants to be housed in local, off-base housing. This was not considered a hardship since Cap St. Jacques had been the premier French seaside resort town, and it was to become one of the favorite in-country 'Rest and Recreation' (R&R) destinations.

OPERATIONS OCTOBER 1962 - 1963

Earlier, in February 1962, the first Army Direct Support Company had been deployed to Nha Trang, and given the impossible mission of supplying and supporting all Army aviation units in Vietnam. In spite of the best intentions and efforts, the support provided was too little and often too late because the distances which separated the supplier from its supported units was just too great. The arrival of the 611th Transportation

Company (Direct Support) was to go a long way towards stabilizing the situation. Based at Vung Tau in October 1962, the 611th assumed responsibility for the southern sector of the country.

As previously mentioned, the unit launched into its mission with enthusiasm, quickly making a name for themselves. In addition to their 'high profile' job of recovering downed aircraft, more often than not, under sporadic enemy sniper fire, the 611th provided 3rd and 4th echelon maintenance support in Vung Tau to all Army aviation units in their sector. They also provided back-up support for the 2nd echelon maintenance performed by the CHFM detachments of the helicopter companies. Soon after they arrived in Vietnam, the 611th relocated their supply section to Saigon, together with personnel from the 339th Transportation Company (DS), to set up and operate the Aviation Supply Point.

A few typical recovery missions of the 611th will suffice to illustrate the unit's operations. On 8 December 1962, one of the 18th Aviation Company's U-1A Otters was ready for take-off at Ca Mau. Both sides of the airstrip were lined with H-21's, on stand down between missions. Colonel Conley recalled that when the 18th's pilot, Captain Wickware, was about to leave the Operations room, someone called out "Hey Wick, watch out for those choppers." To which Captain Wickware responded, to the effect, not to worry, he'd take it right down the middle. Halfway through his take-off run, a sudden gust of crosswind caught the Otter, which caused it to drift into the rows of helicopters, clipping and damaging rotor blades, ending up plowing into a marsh at the end of the runway![7] More work for the 611th, for 8 December 1962 was a day of accidents; in addition to the 18th's Otter and the three H-21's damaged at Ca Mau, two H-21's, one each from the 33rd and 93rd Transportation Companies (Lt Hel), had been damaged upon landing in a LZ, a third grazed one of the downed birds and three more went down with engine failures. All in all nine H-21's and one U-1A were damaged or experienced engine failure in just one day; two of the H-21's had to be destroyed.

1963 started off with a bang at Ap Bac giving the 611th plenty of aircraft to recover as mentioned elsewhere. A sad occasion was to occur on 12 January 1963 when Captain Conley had to locate and recover the 57th's aircraft which had gone down the night before, and on which his friend Captain Don Toth had been a passenger. The accident had been caused by salt water corrosion of the control cables in the rear of the aircraft. The 57th immediately inspected all Shawnee control cables, which required removing them from the aircraft and putting tension on them while visually inspecting the surfaces. This was a monumental task, for more than two hundred cables had to be inspected.[8]

In February the action, in which both Captain Shattuck and CWO Leonard were awarded their second DFC's, occurred during an otherwise routine recovery mission. The two pilots and their crew not only volunteered to carry in 3,500 pounds of ammunition with a full fuel load to a besieged ARVN outpost and deliver it under fire, but carried out 26 ARVN casualties, which, together with the crew, made the total of 30 people carried that day by their Shawnee a record for Vietnam![9]

In March the 1st Aviation Company had a Caribou run off the runway in Rach Gia, referred to as 'Rock Jaw' by many Americans. As was now becoming the norm, the bigger the airplane, the higher the rank of visitor that showed up to 'supervise' the recovery of the taxpayer's property. In the case of the Caribou, Brigadier General Stilwell stopped by to lend a hand.

After June of 1963 the 611th was augmented by two H-37B Mojave's of A Flight, 1st Platoon, 19th Transportation Company. Four Mojave's had been sent to Vietnam, under Captain Charles P. Callaway, two of his aircraft were placed in support of the 611th and two supported the 339th Transportation Company (DS) in Nha

The 611th Transportation Company (DS) insignia is affixed to an O-1 Bird Dog propeller boss, and patches of supported units tacked to the wooden molding underneath, 1963.

Photograph courtesy of Col. Sam Conley, Jr. USA Ret.

A large Transportation Corps emblem and 611th unit designation imbedded in the floor of the unit's mess hall, 1963.

Company sign outside the Orderly Room, Vung Tau 1963.

General Stilwell, wearing a 33rd Transportation Company (Lt Hel) patch on his flight suit, visits the 611th in Vung Tau, March 1963. Behind the General, left to right, unknown, Captain Shattuck, LTC Robertson and CWO Leonard.

Captain Shattuck in the 611th's U-1A Otter, s/n 55-3257. Partially visible is the large eagle marking painted by a very talented unit artist. This unit marking remained on the Otter for only one year, 1963.

CWO John Leonard, in the center, with crew chief SP5 Fred Brown III, and to the right, SP5 Luskey, assistant crew chief in front of the Recovery Platoon H-21C, s/n 56-2134, at Vung Tau in February 1963.

CWO Leonard, Maintenance Officer, poses proudly besides the Recovery Platoon H-21C at Vung Tau in February 1963.

All photographs on this page courtesy of Col. Sam Conley, Jr., USA Ret.

Vietnamese workers building H-21 cradles. The 611th hangar, in the background, was for rotary wing aircraft repair. The small building in the center was a latrine with showers. Vung Tau, February 1963.

The 611th Transportation Company's (DS) machine shop and tool shelters in Vung Tau, February 1963.

Trang. The greater lifting capacity of the H-37B's immediately made itself felt and they were put to good use by the units.[10] Eventually the Mojave's were assigned to the Direct Support Transportation Companies, and A Flight, 1st Platoon, 19th Transportation Company was deactivated.

In its first two years of Vietnam service, over 100 members of the 611th Transportation Company (DS) had received awards, and the unit had recovered 139 downed aircraft, including 54 H-21's and 43 UH-1B's.[11]

The 611th Recovery Team on 12 January 1963 go about the grim business of recovering the bodies of seven Americans lost when an H-21, s/n 55-4220, of the 57th Transportation Company (Lt Hel) fell out of the sky the night before onto an island in the Mekong River.

Captain Samuel G. Conley, Jr. Commander 611th Transportation Company (DS) in February 1963.

AIRCRAFT PAINTING AND MARKINGS

The 611th was assigned three aircraft, a U-1A Otter, s/n 55-3257, a U-6A Beaver, s/n 54-1727, and an H-21C Shawnee, s/n 56-2134. All three were painted in Gloss Olive Drab (FS. 14087) with the Army's standard high visibility markings. The Recovery Platoon's H-21C was quickly 'toned down', having the National Star Insignia and the high visibility markings painted over. Just the last five, '62134', of the serial number was carried on the vertical tail in yellow. This particular H-21C had a fabricated fairing over the nose landing gear, which gave it a distinctive look.

611th Recovery Team about to launch on a recovery mission in February 1963. Team leader SFC Ivon Coleman is standing in the rear row, second from right.

SFC Coleman's Recovery Team on location recovering a downed H-21 in March 1963.

Two of the unit's aircraft carried unique markings during 1963. A large eagle marking was painted just behind the pilot's door of the unit's U-1A Otter by a very talented unit artist. This artistic individual also painted the eagle, holding an ax in each talon, on the Recovery Platoon's H-21C after Ap Bac, with the legend 'Have Axes, Will Travel'.

All photographs on this page courtesy of Col. Sam Conley, Jr., USA Ret.

A remarkable action photograph of the 611th's CWO Carl Flippin chopping the wing off of a crashed 1st Aviation Company CV-2 Caribou at Rach Gia in March 1963.

The Caribou's wing then could be airlifted out by the 611th's Recovery Platoon H-21C, s/n 56-2134.

The 611th Transportation Company (DS) sponsored a theater-wide Maintenance and Commanders conference in May 1963. Brigadier General Stilwell, other commanders and everyone who was anyone in Army aviation maintenance in Vietnam was present.

All photographs on this page courtesy of Col. Sam Conley, Jr., USA Ret.

Another view of attendees, at the back of the room, at the Maintenance and Commanders conference in Vung Tau.

Patch of the Aviation Section, 3rd RRU 1963-1966

Unofficial 'beercan' insignia of the 3rd Radio Research Unit, 1962-1966

Aviation Section, 3rd Radio Research Unit

VIETNAM SERVICE

The 3rd Radio Research Unit gained additional capability when three aircraft were outfitted with ARDF gear in March 1962. The Aviation Section was organized in June 1963. Unit served in Vietnam until inactivated on or about 19 May 1966; the personnel, assets and equipment being redesignated the 509th Radio Research Group on 20 May 1966, per General Order 48, Headquarters, US Army Security Agency.[1]

UNIT NOTES

Part of the National Security Agency's Operation White Oak, dedicated to gathering detailed intelligence on North Vietnam, the 400th U.S. Army Security Agency (USASA) Operations Unit (Provisional) was originally activated in May 1961 and operated under the cover designation of the 3rd Radio Research Unit (RRU). The 400th was renamed the 53rd USASA Special Operations Unit, maintaining the 3rd RRU cover designation, on 20 September 1961.[2]

FACILITIES

Unit was based at Tan Son Nhut in its own compound, which was named "Davis Station" to honor the first unit member, SP4 James T. Davis, who was killed on 22 December 1961 advising the ARVN. Headquarters 3rd RRU, in Saigon, was referred to as "Net Control."[3] Unit eventually forward deployed aircraft to Can Tho, Da Nang and Phu Bai.[4]

OPERATIONS

Airborne Radio Direction Finding (ARDF) gear was hand-carried to Vietnam by Herbert S. Hovey, Jr. of the Army Security Agency and Harold M. Jaffe of the Electronic Warfare Laboratory at Fort Monmouth, New Jersey and installed on three U-6A's that were "locally obtained."[5] These first three RU-6A's became operational in March 1962. The first two RU-8D's and one U-8F were received by the unit in January 1963. It appears that the direction finding equipment performance in the U-8F was unsatisfactory and the aircraft reverted to a liaison role.[6] Seven more RU-6A's, configured in the United States, were taken on charge in 1964, these aircraft being delivered under the project name 'Seven Roses.'[7]

The 3rd RRU both intercepted North Vietnamese and Viet Cong transmissions and recorded them so that they could be immediately forwarded to the National Security Agency (NSA) at Fort Meade, Maryland. The aircraft of the 3rd Radio Research Unit flew radio intercept intelligence gathering missions, basically to eavesdrop on Viet Cong radio traffic. This flying platform for ARDF equipment also allowed Army security personnel, known as direction finding operators (DFO), to use enemy signal frequencies to establish a triangulation and attempt to pinpoint the location of secret enemy radio transmitters.[8] It was just such a mission in late December 1962 which led to the action at Ap Bac on 2 January 1963.

The Aviation Section, 3rd Radio Research Unit was officially organized in June 1963.[9] The unit later received additional RU-6A's and RU-8D 'Checkmate' aircraft, and 'Winebottle' and 'Cefish Person' systems equipped aircraft; additionally one example of a RCV-2B 'Pathfinder' Caribou, s/n 62-4147, was received in January 1966 prior to the unit's deactivation.

By 1966 the Aviation 'Section' had grown to approximate battalion size and there exists a second, unofficial unit patch identical to the one illustrated except that it had BATTALION written where SECTION was previously located.[10]

AIRCRAFT PAINTING AND MARKINGS

The RU-6A's (L-20A) and RU-8D's (RL-23D) of the Aviation Section, 3rd RRU carried the Gloss Olive Drab (FS. 14087) overall paint scheme with the Army's standard yellow and white high visibility markings. The unit maintained their aircraft in an extremely clean condition, which, given the nature of their mission to 'bore holes in the sky', was not too difficult to do.

After 1965, ARDF aircraft were received in the new Army camouflage paint scheme of overall Flat Olive Drab (FS. 34087) with black low visibility lettering, markings and serial number. Aircraft already serving with the unit would have been repainted in the new scheme as time and scheduled maintenance permitted.

Due to the security conscious, low profile nature of the 3rd RRU Aviation Section information about and photographs of their aircraft have not been easy to locate. It is not known if the Aviation Section carried the unit's insignia or a unit marking on their aircraft. At least one, if not more, RU-6A carried a personal marking on the nose of the aircraft.

Almost certainly one of the first three aircraft to be fitted with ARDF gear was RU-6A, s/n 55-682, Vung Tau in January 1963. Photograph courtesy of Col. Sam Conley, Jr., USA Ret.

RU-8D, s/n 58-3075, of the Aviation Section, 3rd RRU at Tan Son Nhut photographed in February 1964. USAF Photograph

Right and Profile Below: **RU-6A, s/n 57-6151, of the Aviation Section, 3rd RRU at Tan Son Nhut in February 1964. Also visible is a MACV-MAAG U-6A, s/n 51-16839 (formerly the floatplane), and, in the background, a 61st Aviation Co. (CV-2) Caribou, s/n 61-2385. The RU-6A is painted in overall Gloss Olive Drab (FS. 14087) with the Army's standard high visibility markings and a black antiglare panel on the upper cowling. The nickname 'Slow Poke' is scripted in yellow on the nose.**

USAF Photograph

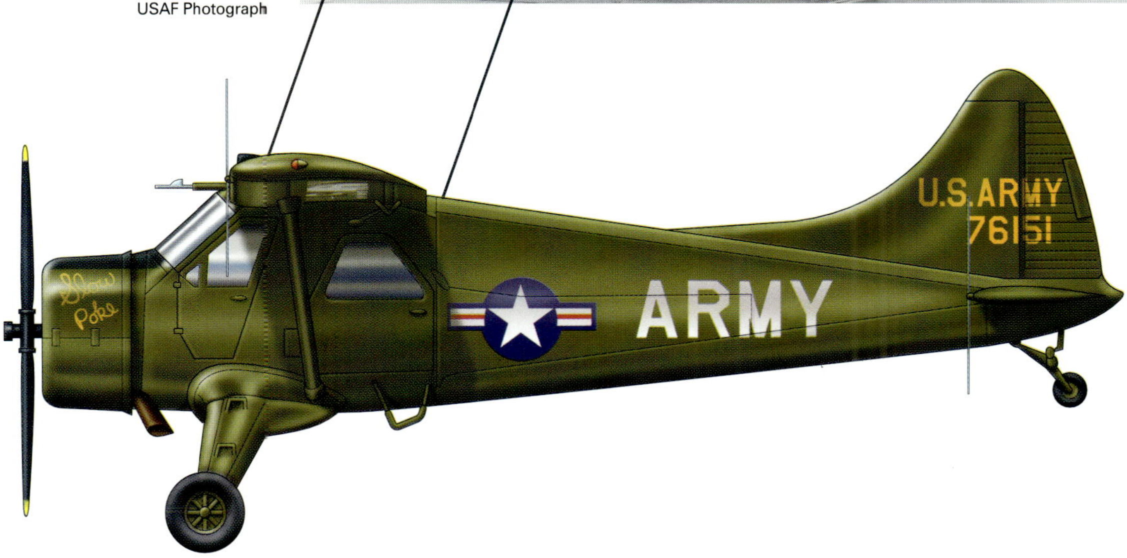

1st Aviation Company (CV-2)

UNIT NICKNAME "PATHFINDER" 1962 - 1963

70TH TRANSPORTATION DETACHMENT (FIELD MAINTENANCE)

DETACHMENT MOTTO "GREMLUS PUGNACUS" 1963

1st Aviation Company patch, US manufacture (Fixed Wing-Tactical Transport).
Patch courtesy of CW4 Don Joyce, USA Ret.

1st Aviation Company patch (Fixed Wing-Light Transport).
Patch courtesy of Chris Miller.

70th Trans. Det. (Field Maintenance) patch. The wording "Gremlus Pugnacus" may have appeared on patch.
Provisional drawing, artwork by Kevin Lyles.

VIETNAM SERVICE

The 1st Platoon, 1st Aviation Company arrived in Vietnam in July 1962 with its 8 aircraft. The remainder of the unit's 10 aircraft deployed to Vung Tau in December 1962. On 7 December 1963 the 1st Aviation Company left Vung Tau and redeployed to the United States. Redesignated as the1st Army Security Company (Aviation) in December 1966, unit returned to Vietnam in June 1967 as 1st Radio Research Company (Aviation).

COMMANDERS, NOTABLE UNIT MEMBERS, TRADITIONS

The 1st Aviation Company (FW-TT) (CV-2) had been formed at Fort Benning, Georgia. In 1961 Major William G. Hooks had assumed command and successfully lead the company through many firsts for Army aviation. Major Hooks deployed the 1st Aviation Company in Operation 'Big Moose', a flight from Georgia to Korat, Thailand, in May 1962 in support of Joint Task Force 116.

The unit quickly made a name for itself in Vietnam. In April 1963 Major Morgan H. Mathews assumed command of the unit.[1] The last unit commander who led the unit in another first for Army aviation, the return flight to Ft. Benning, Georgia was Captain Kenneth Blake, Jr.[2]

Unit nickname "Pathfinder" was acquired due to the unit's pioneering and trailblazing role.

The 70th Transportation Detachment (Field Maintenance) was commanded by Captain George A. Brown, the unit being attached to the 1st Aviation Company in May 1963.

The 70th Trans. Det. wore red hats, with their Detachment designation embroidered on the front of the hat in yellow.

FACILITIES

The unit was based in Vung Tau, which provided better than average housing and mess facilities, and, of course, exceptional recreational opportunities for the unit. Two aircraft sections had operated out of Da Nang and Nha Trang, whose facilities were considered more than adequate.

OPERATIONS

The unit, under Major Hooks, closed at Korat, Thailand with 18 CV-2 Caribou after flying out from Ft. Benning, Georgia between 26 May and 21 June 1962. The aircraft were supposed to have a low profile, especially since the Army had been turned down on deploying a Caribou unit to Vietnam in December 1961 by the Commander-In-Chief, Pacific (CINVPAC) and the USAF were loudly denouncing the Army's use of the Caribou.

The 1st Platoon, with 8 aircraft, was sent in July 1962 to support the three Corps areas, 2 aircraft to Da Nang, I Corps, 2 aircraft to Nha Trang, II Corps and 4 aircraft probably based at Tan Son Nhut, III Corps. The Caribou's of the 1st Platoon immediately went to work supporting U.S. Special Forces and other Army units in Vietnam.[3]

In December 1962 the remaining aircraft of the 1st Aviation Company were ordered to close at Vung Tau. By January 1963 the unit had been integrated into the planning at Corps level. The unit provided a regularly scheduled service, often referred to as "milk runs", throughout Vietnam, and while these runs may not have been as exciting as flying armed helicopters, they provided a much needed service. Lt. Col. Mertel described a typical day in the life of a 1st Aviation Company Caribou crew.

"One such run is the one serving the IV Corps area on a regular schedule, 6 days a week. A typical day starts at 0730 at Tan Son Nhut, Saigon International airport. The Caribou flown by CWO's Phillip Crossan, pilot, and William Everhart, co-pilot, and crewchief SP5 Henry W. King departed on schedule with a full load of U.S. military personnel, Vietnamese military and civilian personnel, U.S. Mail, motion pictures, baggage, critical supplies, etc.

The first stop is at one of the MAAG Detachments. An easy strip, 1850 feet long, light gravel on sod, rice paddies on both sides. Here a change of passengers; some get off, some on. Mail is exchanged, also motion picture film. Next stop, 12 minutes later, is a nice long hard surfaced runway, 2,400 feet. A group of soldiers meet the flight to exchange news, to bid farewell to a rotatee who is beginning his journey back to the U.S. A few light-hearted comments to a buddy off for R&R to Hong Kong or Bangkok. Off again for a 20 minute flight, this time landing on a 3,000 foot sod strip - slippery as glass when wet. A similar exchange of passengers, parcels, mail and packages and take-off again.

Another gravel strip, this time only 1,500 feet usable on a 3,300 foot runway. No problem for the Caribou, runway about half utilized on landing, one third on take-off. Ten minutes later to the next stop - 1,300 feet of hard surface. A short field landing without any difficulty. An almost complete change of passengers at this station, except for rotation and R&R personnel. Two very tired Infantry 1st Lieutenants and a Sergeant get aboard, returning to Saigon for a few days of well earned rest after several months in the field with Vietnamese Army units! As soon as the Caribou is airborne, all three are asleep. Fifteen minutes later another strip followed a few minutes later by yet another. A high approach this time to the field of a major unit command post. From 1,500 feet above the ground, power approach at 65 knots, not because the field is short, but because a Viet Cong sniper may fire at you if you "drag one in." In addition a good crosswind on this one. A crosswind is most common in the Republic of Vietnam, and the wind is always off one wing and at a pretty strong velocity. Off again, two more stops and back to Saigon."[4]

These regularly scheduled runs were extremely popular with the troops, one photograph by Richard Tregaskis in his book 'Vietnam Diary', shows the obvious happiness of a group from the 93rd Transportation Company (Lt Hel) getting ready to board a Caribou at Soc Trang heading home.[5] However these runs could be taxing on the crew, one aviator with the 1st Aviation Company made 26 landings and 26 take-offs on his 5 1/2 hour "milk run."[6]

The 1st Aviation Company (CV-2) compiled an outstanding record while in Vietnam flying an appreciated but unheralded mission. Flying in support of U.S. Special Forces outposts in the Central Highlands the airstrips were often short and narrow, sometimes carved out of the jungle. Using these remote airstrips was a high risk since the enemy often attempted to destroy any aircraft on the ground and the area around the posts were not secured. When flying in support of Vietnamese Army outposts in I Corps the cargo could include rice, live chickens, goats and pigs in wicker baskets, in addition to personnel and equipment.[7]

The 1st Aviation Company (CV-2) was redeployed to the United States, as part of a token force drawdown, in early December 1963.

AIRCRAFT PAINTING AND MARKINGS

Most Caribou of the 1st Aviation Company (CV-2) carried their unit insignia on their tails, under the aircraft serial numbers, while at Ft. Benning, Georgia.[8] However when the 1st Aviation Company deployed to Korat, Thailand in May 1962, in keeping with the unit maintaining a low profile, the insignia had been removed from the aircraft. Aircraft arrived in Vietnam painted in Gloss Olive Drab (FS. 14087) with the Army's standard high visibility markings.

The 1st Aviation Company's aircraft were kept clean, and plain, to attract the least amount of attention. While in Vietnam the unit removed the propeller spinners from their aircraft. No unit markings or personal markings were observed on any of their aircraft through 1963, except one. The one aircraft, serial number unknown, which carried a nose marking of a 'large cluster of red cherries' was the only aircraft in the 1st Aviation Company which had not been hit by enemy fire during the unit's service in Vietnam. This marking was present, and written about, when the unit was redeploying to Ft. Benning, Georgia from Vietnam in early December 1963.[9]

CV-2A Caribous, s/n 61-2387 and 61-2391 visible, of the 1st Aviation Company in Korat, Thailand in 1962 prior to deploying to Vietnam. US Army Photograph

Another view of CV-2A, s/n 61-2391, being refueled at Korat, Thailand in 1962. US Army Photograph

A downed Caribou, s/n 61-2388, is given an assist by an A Flight, 1st Platoon, 19th Trans. Co. H-37B, s/n 55-636. Photograph courtesy of Col. Sam Conley, Jr., USA Ret.

A 1st Aviation Company Caribou, s/n 61-2387, at Vung Tau. The U-1A, s/n 58-1713, of the 18th Avn. Co. had crashed at Ca Mau on 8 December 1962. Photograph courtesy of Col. Sam Conley, Jr., USA Ret.

Major William G. Hooks, commander of the 1st Aviation Company (CV-2) in Vietnam, center, stands with General Stilwell and Lieutenant Colonel Robertson in front of the General's Bird Dog, possibly s/n 51-17450, at Vung Tau in March 1963. Photograph courtesy of Col. Sam Conley, Jr., USA Ret.

Major Morgan H. Mathews, the second unit commander, in an informal pose in the 611th Trans. Co. (DS) Operations building in April 1963. Photograph courtesy of Col. Sam Conley, Jr., USA Ret.

Standing in front of a 1st Aviation Company Caribou are, left to right, Captain K. Blake, Jr. XO (and later CO), Captain H. Chambers, Lieutenant R. Rodriguez and Lieutenant M. Brooke in July 1963.

Photograph courtesy of Mel L. Brooke

1st Aviation Company's area in November 1963. Note the unit's insignia both on the building and the M-151 of the 70th Trans. Det. The wheel cover on the Dentist's vehicle is of interest as are the early issue canvas and leather footwear worn by unit members.

Photograph courtesy of Mel L. Brooke

A 1st Aviation Company CV-2A, s/n 61-2385, supporting the UTT with fuel at Phi Truong Buu Son during their flight from Tan Son Nhut to Da Nang in June 1963. One of the UTT's UH-1B's, s/n 62-1949, can be seen in the background.

Photograph courtesy of Col. Robert L. Hoffman, USA Ret.

Another perspective of the 1st Aviation Company (CV-2) refueling the UTT. UH-1B s/n 62-1946, awaits its turn at Phi Truong Buu Son in June 1963.

Photograph courtesy of Col. Robert L. Hoffman, USA Ret.

Only known group photograph of the 1st Aviation Company (CV-2) in Vietnam, July 1963. Major Morgan Mathews, Commanding Officer, is seated in the center of the first row. The commander of the 70th Transportation Detachment (Field Maintenance), Captain George A. Brown, is wearing the red cap.

Photograph courtesy of Mel L. Brooke

MAAG Vietnam emblem.

MACV emblem.

Neither a MAAG, nor MACV-MAAG Flight Detachment patch has been identified.

VIETNAM SERVICE

The original MAAG Flight Detachment had been formed in October 1954 and became the MACV-MAAG Flight Detachment in February 1962. Unit was renamed the MACV Flight Detachment on 15 May 1964 when the Military Assistance Advisory Group, Vietnam was formally dissolved.

COMMANDERS, NOTABLE UNIT MEMBERS, TRADITIONS

The Flight Detachment initially came under the command of the MAAG Aviation Officer and, later, after the activation of the Military Assistance Command, Vietnam, it is believed the Senior Aviation Officer, MACV.

MACV-MAAG Flight Detachment

The Assistant Army Attache to Laos, then Major Leland F. Wilhelm described his tour in Vientiane, Laos from an Army aviator's point of view in his article "The Laos Story." The MAAG aircraft assigned to Laos, L-23 s/n 53-6183, was painted in the red and white scheme described below.

Aircraft was ferried into Vientiane, Laos from Tokyo by Major Wilhelm.[1]

In 1962 to 1963 the MAAG Aviation Officer was Colonel Delbert Bristol, who blazed the trail for Army aviation in Vietnam, and who was succeeded in March 1963 by Colonel Guy Jones. The first MACV Aviation Officer was Lieutenant Colonel Otto Huebner who was replaced, upon his departure after his one year tour of duty, by Lieutenant Colonel Arne Eliasson.[2]

FACILITIES

The MAAG and, later the MACV-MAAG, Flight Detachment was located at Tan Son Nhut airbase, with quarters in downtown Saigon. At least one aircraft was based in Vientiane, Laos to support the Army's advisory team in that country. The unit generally had four or five aircraft assigned at any one time which were strictly used as VIP aircraft. From 1961 through 1962 the Flight Detachment operated a L-20A (U-6A) float plane, s/n 51-16839, which was moored in the Saigon River. This particular aircraft had the distinction of being the only Army float plane in Vietnam! The float plane was converted back to a wheeled configuration in 1963, having towed one too many waterskiers on the Saigon River![3]

OPERATIONS

The original MAAG Flight Detachment provided air transportation to the Military Assistance Advisory Group, Indochina in support of their mission. This required flights into Laos and Cambodia since these countries were part of MAAG, Indochina. An aircraft was subsequently based in Laos.

An extract from the Command Aircraft Company (CAC), the direct descendant of the original MAAG Flight Detachment, states that the MAAG Flight Detachment also operated one, or more C-45 aircraft, in addition to the L-20A Beavers and L-23D Seminoles.[4]

On 8 February 1962, when the Military Assisstance Command, Vietnam was officially activated, the unit became the MACV - MAAG Flight Detachment. Typically the aircraft were reserved for the MACV and MAAG commanders, General Harkins, who used the L-23D almost exclusively, and General Timmes, chief of the advisory group, as well as to transport senior U.S. Government officials on visits to Vietnam and high ranking U.S. Military Commanders. Depending on availability, and the mission requirements, others had access to the aircraft from time to time. The early H-21 units had 'borrowed' the Flight Detachment's L-20A's to use as a higher flying directional controller during the first combat assault missions. Subsequently the H-21 units had received their own L-19's (TO-1D) and the Flight Detachment's VIP aircraft reverted to their more sedate missions.

Known MACV - MAAG Flight Detachment aircraft serial numbers L-20A (U-6A) 51-16839 and 53-

The MACV-MAAG Flight Detachment U-6A float plane, s/n 51-16839, being beached at Vung Tau in late 1962 after a rough crash landing. Photograph courtesy of Col. Sam Conley, Jr., USA Ret.

The MACV-MAAG Flight Detachment U-6A, s/n 51-16839, with replacement floats flown in from Canada, in the fixed wing repair hangar of the 611th Transportation Company (DS) at Vung Tau in late 1962. The American flag decal is visible on the tail.
Photograph courtesy of Col. Sam Conley, Jr., USA Ret.

7959; L-23D (U-8D) 57-6033, and the MAAG aircraft, L-23 53-6183, was based at Vientiane, Laos. By early 1963 the Flight Detachment had gained at least one L-23F.

AIRCRAFT PAINTING AND MARKINGS

The aircraft of the MACV - MAAG Flight Detachment were painted in the color scheme designated 'Arctic, Desert and Jungle Markings' which consisted of a bright red and white pattern. The intent of such a bright paint scheme was to facilitate rescue and recovery if an aircraft should crash in the jungle. The colors were a Gloss Red (FS. 11136) and Gloss White (FS. 17875). As the aircraft were expected to fly into neighboring states, and even Thailand, American flag decals were carried on the tails.[5]

The Flight Detachment's L-23F, s/n believed to be 62-3867, in a gloss olive drab and white scheme photographed at Vung Tau in February 1963.
Photograph courtesy of Col. Sam Conley, Jr., USA Ret.

Above and Profile Left: U-8D, s/n 57-6033, of the MACV-MAAG Flight Detachment in the Gloss Red and Gloss White scheme in early 1964. Also visible U-6A, s/n 53-7959. Note the position of the American flag decals used on all MACV-MAAG aircraft. USAF Photograph

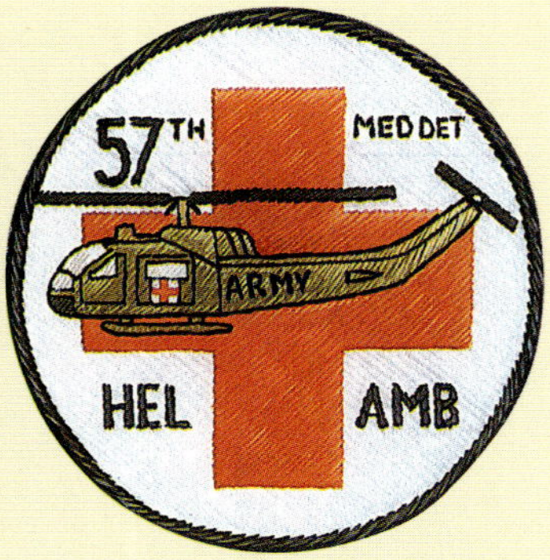

Rare, first patch of the 57th Med. Det. (Hel Amb) 1962.
Artwork by Kevin Lyles.

Second patch adopted after unit took "DUST OFF" call sign in 1963.
Patch courtesy of Chris Miller.

57th Medical Detachment (Helicopter Ambulance)

UNIT CALL SIGN 'DUST OFF' 1963-1973

VIETNAM SERVICE

The 57th Medical Detachment (Helicopter Ambulance) arrived in Vietnam in April 1962. This legendary unit served with great distinction throughout the war. The 57th was the last Helicopter Ambulance detachment to serve in Vietnam, turning in their helicopters on 14 March 1973.[1]

COMMANDERS, NOTABLE UNIT MEMBERS, TRADITIONS

The first commander of the 57th Medical Detachment (Hel Amb) was Captain John Temperelli, Jr. Captain Temperelli oversaw the unit's deployment to Vietnam and endured the numerous frustrations attendant with the introduction of a new unit, concept and helicopter to service. Captain Temperelli however was a fighter and, in what was probably the fight of his career, from May 1962 to February 1963 he fought continuously to define his mission to MACV and USASG-V, to keep his aircraft and to 'go to where the action was.' In retrospect he was the 'father' of the concept that was to become famous in Vietnam.

The second commander of the 57th Medical Detachment (Hel Amb), who took over the unit in February 1963, was Major Lloyd Spencer. Major Spencer was responsible for picking the call sign 'DUST OFF' when he had determined that the unit needed its own call sign. Previously all communications went through Vietnamese radio channels, which made it impossible for supported American units or advisors to contact the 57th Medical Detachment.

Major Spencer visited the Naval Support Activity in Saigon, the agency which in 1963 had responsibility for assigning communications frequencies and call signs, and, in the Signal Operations Instructions (SOI) code book, listing all call signs and radio frequencies in Vietnam, he chose the unused call sign DUST OFF. Later, in mid 1963, the National Security Agency (NSA) initiated a review of all call signs in Vietnam, and attempted to reassign the 'DUST OFF' call sign. The 57th Medical Detachment (Hel Amb) took on the NSA and with the increased recognition, requirement for and popularity of his unit, refused to give up their call sign. The NSA had to back down and DUST OFF remained the 57th's call sign.[2] Major Spencer commanded the unit until 11 January 1964, when one of the most famous DUST OFF commanders, Major Charles L. Kelly, assumed command.

In 1962 the unit had a small, pet spider monkey, named 'Huey', as their mascot.[3] Generally these small monkeys were " most friendly and really like to grab you. They will hide behind a corner; then when some unsuspecting soul appears, come bounding out, run up his leg, and perch on his shoulder. This is a bit startling the first time it happens."[4]

FACILITIES

Nobody was prepared for, or even had been notified of, the arrival of the 57th Medical Detachment (Hel Amb) in Nha Trang. Captain Temperelli reported to the US Army 8th Field Hospital which had recently set up in Nha Trang. A small helicopter ambulance unit was not a high priority, and the commander of the 8th Field Hospital assigned the unit some tents on concrete pads. The unit made do, but the facilities were not the best. When the unit finally got their five UH-1A's in May 1962, two of them were sent up to Qui Nhon to support northern II Corps and I Corps.

The 57th Medical Detachment (Hel Amb) remained headquartered in Nha Trang until 16 January 1963 when the unit was redeployed to Tan Son Nhut Air Base.[5] The facilities that the unit occupied improved considerably and, of course, there was always Saigon a short taxi ride away. In April 1963 two aircraft, of the unit's total of five, were reassigned to Pleiku in the Central Highlands.

OPERATIONS

UH-1A, s/n 58-2081, of the 57th Medical Detachment (Hel Amb) at Nha Trang in May 1962. The 'last two' of the serial number is repeated in white on the nose red cross, a unit practice on their UH-1A's. The UH-1A Otter, s/n 58-1695, in the background belongs to the 18th Aviation Company.

Photograph by Bill Hardy, Wayne Mutza collection

The advance party of the 57th Medical Detachment (Hel Amb) arrived in Nha Trang on 26 April 1962. In early May the five UH-1A's assigned to the unit had arrived in Saigon and Captain Temperelli, the unit's Commanding Officer and one of his pilots went to pick up two of them. Getting the two helicopters to Nha Trang was an adventure in itself since there were no facilities enroute in which to refuel. Eventually the unit got its five helicopters, two of them being sent to Qui Nhon, even though there was no JP-4 fuel for the Hueys in Qui Nhon, and the unit had to use a field expedient solution to the problem, by placing one of their two 1,200 gallon fuel storage tanks in Qui Nhon and flying fuel up to it in 55 gallon drums.[6]

The first casualty picked up was a wounded Vietnamese, in spite of a MAAG policy against medevacing the ARVN. However Nha Trang was far from the action and the unit was under utilized. As combat units, such as the UTT, received the new UH-1B model Huey there was increased demand on spare parts and the low utilization of the 57th Medical Detachment put pressure on the U.S. Army Support Command, Vietnam to take the units aircraft. The situation erupted in November 1962 when the ARVN planned a major operation and the starter generators of the unit's Hueys were requisitioned. While Captain Temperelli fought this decision all the way to General Stilwell, he lost and from mid November to mid December the unit was grounded for lack of parts. To add to the unit's woes, only one starter generator was returned and the unit started the New Year in 1963 with one operational helicopter.

However the action at Ap Bac rewrote the equation, and opinions, at the highest levels, were changed. Captain Temperelli was ordered to relocate his unit from Nha Trang to Tan Son Nhut. As the relocation occurred, Major Spencer assumed command of the unit in February and the UH-1A's were replaced by five brand new UH-1B's in early March 1963 which had just been delivered by the USNS Brostrom.

By the end of 1963, the unit was separated between three aircraft based at Tan Son Nhut and two in Pleiku, and had flown 1,485 missions and evacuated 1,972 wounded.[7]

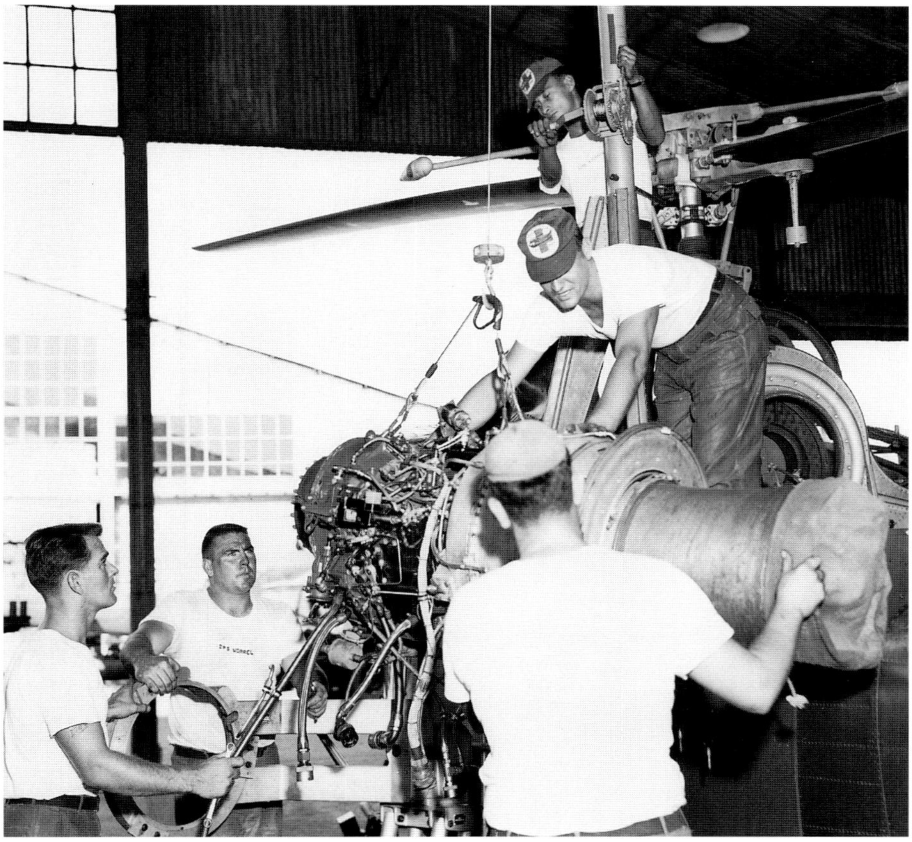

57th Medical Detachment (Hel Amb) maintenance crews remove an engine from a UH-1A on 28 November 1962. The unit was effectively grounded until January 1963. Note the first unit patch, before the "DUST OFF" addition, being worn on the utility caps of two individuals.

US Army Photograph

AIRCRAFT PAINTING AND MARKINGS

The unit's five UH-1A's arrived in Vietnam in Gloss Olive Drab (FS. 14087) and the Army's standard high visibility markings. In addition they carried the authorized medical insignia, red crosses on white squares, in three positions as required in 1962, one located on the nose and two centered on the aircraft's cargo doors.

Two, or more, of the unit's aircraft, one of them was UH-1A s/n 58-3035, carried the marking AMBULANCE stenciled in white on the cargo door immediately above the window. The unit's aircraft generally carried the 'last two' of the serial number, such as '81' on the nose of UH-1A s/n 58-2081, in white centered within the nose red cross.[8]

The unit received five brand new UH-1B's to replace their UH-1A's in March 1963. These UH-1B's also carried the Army's standard, gloss paint scheme and high visibility markings, with the addition of red crosses on white squares. No unit insignia, or other unit markings, were observed on the 57th Medical Detachment's UH-1B's in 1963.

UH-1A's, s/n 58-2081 and 58-3035, reportedly the first two 57th Medical Detachment aircraft in Vietnam, and, in all probability, the first Bell UH-1A's in Vietnam in May 1962. The second aircraft has 'AMBULANCE' in white lettering over the cargo door window. In the background can be seen an H-21 of the 8th Transportation Company (Lt Hel).

Photograph by Bill Hardy, Wayne Mutza collection

UH-1B, s/n 62-1887, of the 57th Medical Detachment (Hel Amb) landing to pickup a wounded Vietnamese soldier in May 1963.

US Army Photograph

UH-1B, s/n 62-1889, of the 57th Medical detachment (Hel Amb) picking up wounded in October 1963. Aircraft in overall gloss olive drab and full high visibility markings.

US Army Photograph

UH-1A, s/n 58-2081, with the 57th Medical Detachment (Helicopter Ambulance) in April 1962. Aircraft painted in Gloss Olive Drab (FS. 14087) with standard high visibility markings, three position red cross markings and the 'last two' of the serial number in the center of the nose red cross medical insignia.

A Flight, 1st Platoon, 19th Transportation Company (H-37)

UNIT NICKNAME "WINGS OF HELL" - 1963

VIETNAM SERVICE

The unit deployed from Korea to Vietnam, arriving in early June 1963. Unit believed to have been inactivated in 1964, personnel, assets and equipment being distributed to the two Direct Support Companies.

COMMANDERS, NOTABLE UNIT MEMBERS AND TRADITIONS

The unit deployed from Korea under the command of Captain Charles P. Callaway. Captain Callaway would have a long association with the Mojave. In the course of his Army career he took delivery of the second production H-37, introduced the aircraft to combat in Vietnam, and subsequently piloted the last, active duty H-37 on its final flight to demilitarization in 1971.[1] After service in Vietnam, Captain Callaway returned to the United States on 17 October 1963.

At first A Flight was TDY from Korea, and under the operational control of ACTIV. The "initial effort was dedicated to designing and proving out rigging gear and ancillary equipment to allow the safe evacuation of all types of disabled aircraft."[2] As a result of this testing, later direct support maintenance companies would be given some form of medium lift recovery capability, in the form of H-37's and later the Chinook.

To build esprit the unit wore (Transportation Corps) brick red caps with an unofficial cap insignia.

FACILITIES

Unit was first attached to the 611th Transportation Company (DS) in Vung Tau for rations and quarters, and this attachment was later changed to the 330th Transportation Company (GS) also in Vung Tau. The two aircraft and their support crew sent to assist the 339th Transportation Company (DS) in Nha Trang also shared the billet's of the supported unit.

OPERATIONS

Four H-37B's, serial numbers 55-623, 55-625, 55-627 and 55-636, of A Flight, 1st Platoon, 19th Transportation Company were loaded on the USNS Brostrom on 25 May 1963 in Inchon Harbor, Korea.[3] Setting sail for Vietnam, the ship docked in Saigon on 3 June 1963. At first the unit was based in Vung Tau, and attached to the 611th Transportation Com-

19th Transportation Company (H-37) patch.
Patch courtesy of Chris Miller.

"Wings of Hell" unofficial A Flight, 1st Platoon cap insignia, 1963.
Provisional drawing, artwork courtesy of Kevin Lyles.

One of the unit's H-37B's being loaded into the USNS Brostrom in Inchon harbor, Korea on 25 May 1963. The H-37's had already received their white 'conspicuity' markings for Vietnam operations.
Photograph courtesy of LTC Charles Callaway, USA Ret.

pany (DS), conducting their first operational mission on 10 June 1963.

Soon after, in late July, ACTIV determined to provide support to all Corps areas, and the unit was separated, two aircraft, 55-623 and 55-625, remaining at Vung Tau supporting the 611th Transportation Company (DS), and two, 55-627 and 55-636, were flown to Nha Trang to support the 339th Transportation Company (DS).

The arrival of the H-37B's gave a tremendous boost to the aircraft recovery effort, as they provided a far greater lifting capability than the Direct Support companies' H-21C's. The H-37B could lift out, with minimum disassembly, the H-21 and could transport the major components of a CV-2. Another plus, since the Viet Cong often would contest recovery efforts, was the H-37B's ability to absorb battle damage, far better than most Army aircraft.

Then Captain Callaway recalls that his unit averaged "over 100 hours per ship per month prowling around RVN supporting air operations from the Delta to the northern border. In her own right she (the H-37) was truly an unheralded heroine in every sense of the word. Just ask any pilot or crewman."[4]

AIRCRAFT PAINTING AND MARKINGS

When alerted for movement to Vietnam, A Flight prepared its aircraft by painting over the high visibility markings and adding the white stripe on the top of the aircraft. They were the first unit to arrive in Vietnam in the low visibility paint scheme that the Army would later, in 1965, call a camouflage paint scheme and require for all aircraft in combat or in a combat support role. The H-37B's of A Flight, 1st Platoon, 19th Transportation Company arrived in Vietnam on 3 June 1963 painted in Gloss Olive Drab (FS. 14087) with low-visibility markings, just carrying the aircraft serial number painted in yellow on the aft fuselage, and the six inch wide white conspicuity stripes on the top of the aircraft. Personal markings may have been carried on one, or more, aircraft generally painted to the right of the cargo door.

H-37B, s/n 55-623, of A Flight, 1st Platoon, 19th Transportation Company recovers a UH-1 at Vung Tau, July 1963.
Photograph courtesy of Col. Sam Conley, Jr., USA Ret.

Below: **CWO-4 Bill Roundy in one of the unit's H-37B's on route from Vung Tau to Rach Gia on 16 June 1963. CWO Roundy was the unit's standardization pilot.**
Photograph courtesy of LTC Charles Callaway, USA Ret.

H-37B, s/n 55-636, of A Flight lifts a damaged H-21C, s/n 52-8617, of the 57th Transportation Company (Lt Hel), mid 1963.
US Army Photograph

Assisting the 611th Transportation Company (DS) recover a 1st Aviation Company's (CV-2) downed Caribou, the H-37B's significantly expanded the Army 's recovery effort in Vietnam.
Photograph courtesy of Col. Sam Conley, Jr., USA Ret.

Individuals involved in the recovery of a CV-2 Caribou at Ha Tien are, left to right, CWO Flippin, 611th Transportation Company (DS), unknown, the Special Forces A Team commander whose unit had secured the crash site, Captain Callaway, Commanding Officer and Lieutenant Hardy, A Flight, 1st Platoon, 19th Transportation Company and unknown. Both Captain Callaway and Lieutenant Hardy are wearing the 'Wings of Hell' cap insignia. Date approximately late July 1963. Colonel Callaway shared the following regarding this photograph "of a bunch of guys inspecting a hole in the ground"; the Caribou crashed when the "aircraft lost an engine on takeoff and the pilot applied his emergency procedures but pulled off the good engine giving full power to the bad. The 'Bou quickly stopped flying and dipped the left wing into the ground breaking it off. The aircraft then spun around 180 degrees and came to rest. Our mission was to remove the nose section which was undamaged to repair another Caribou that had landed with the nose gear retracted. We were then to haul the remainder of the aircraft offshore and deep six it so that the VC could not make any use of the hulk.

The right wing was completely intact and undamaged. This would make the sling load rather unwieldy so someone decided to remove the wing. Of course removing a wing by drilling out the rivets would have taken many hours, so someone, whose name won't be mentioned, came up with the brilliant thought that the wing could be cut with an explosive charge.

The A Team had a young demolitions expert who was more than eager to demonstrate his skill with C-4. He dragged out his 'How To Blow Things Up Manual', measured the width and depth of the wing root, and quickly computed the size of the charge and went to work fashioning a rather hefty rope of C-4 which he looped around the wing root and tamped with the old faithful green tape which has held the Army together for years. He fused this concoction and ordering the assembled group back 10 or 15 feet, he yelled "Fire in the Hole." Not being as forgetful then as I am now, I remembered a demolition class at the Benning School for Boys. Something rang a bell about the size of charges for solid and hollow objects being vastly different.

Being a devout chicken, I elected to start running as fast as I could from the site, and upon hearing a resounding BOOM, dove head first and attempted to dig a hasty fox hole with my oversized nose. Looking up, I saw bits and pieces of Caribou landing 30 to 40 feet past me. Looking at the CV-2, the wing was blown off as was the wing center section and the fuselage had been relocated approximately 20 feet from its previous resting place now marked by the hole being examined. There was also a 3/4 ton truck sized hole in the fuselage. That is the background of the picture" above. [5]

Photograph courtesy of Col. Sam Conley, Jr., USA Ret.

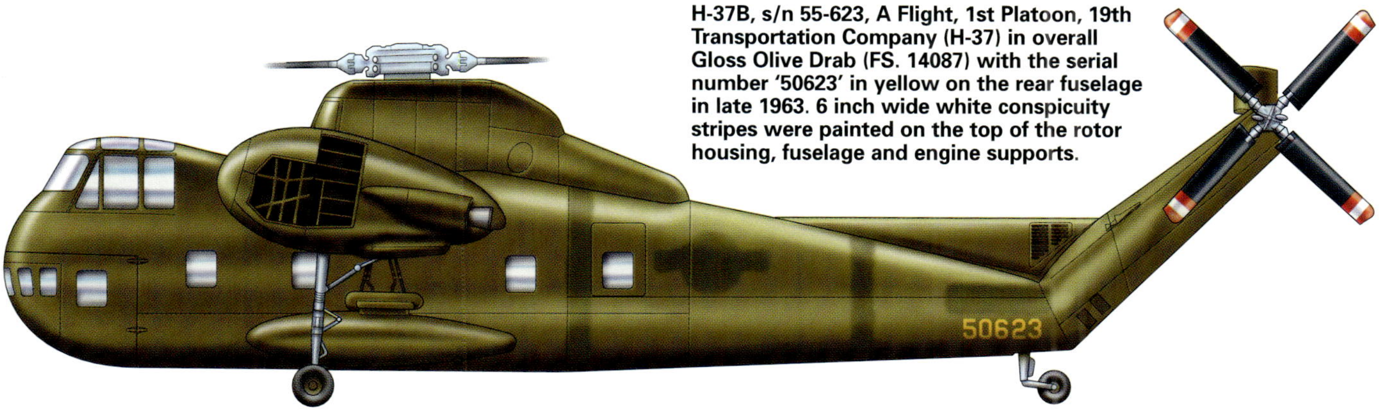

H-37B, s/n 55-623, A Flight, 1st Platoon, 19th Transportation Company (H-37) in overall Gloss Olive Drab (FS. 14087) with the serial number '50623' in yellow on the rear fuselage in late 1963. 6 inch wide white conspicuity stripes were painted on the top of the rotor housing, fuselage and engine supports.

UH-1A Flight Section
45th Transportation Battalion (Trans. Acft.)

UNIT NICKNAME "IROQUOIS" 1962-1963

Patch of the UH-1 Flight Section/Combat Support Section 1962-1963.
Patch courtesy of Chris Miller

Headquarters of the UH-1 Flight Section/Combat Support Section at Tan son Nhut circa mid 1963.
Photograph courtesy of Col. Robert L. Hoffman, USA Ret.

The 'Iroquois Village' compound; note 'The Wigwam' sign in the background which may have been a unit club.
Photograph courtesy of Col. Robert L. Hoffman, USA Ret.

VIETNAM SERVICE

The UH-1A Flight Section arrived in Vietnam in July 1962 and was attached to the 45th Transportation Battalion (Trans. Acft.). Unit served until inactivated in September 1963. All personnel, assets and equipment were transferred to the successor unit, the 145th Airlift Platoon.

UNIT NOTES

This small flight section was based at Tan Son Nhut, Saigon, having its own compound known as "Iroquois Village". The unit supported MAAG advisors throughout all Corps areas in 1962 and 1963. One of the UH-1A Flight Section's first commanders was Captain Merle Bounds.[1] After delivery to the unit of UH-1B's in March 1963, the UH-1A's were soon phased out, and the unit's designation may have then changed to that of UH-1 Flight Section and/or Combat Support Section.[2]

The first two Army aviators to be recommended for the award of the Silver Star in Vietnam were Captains Ray Vining and Joseph A. Josh, of the UH-1A Flight Section, for their actions during Lam Son II in August 1962 in I Corps. They had flown to the aid of the Commanding Officer of the 93rd Transportation Company (Lt Hel), Major James E. Gray, Jr. who's aircraft had been shot down, one of two aircraft downed in the landing zone. The two UH-1A Flight Section pilots braved heavy enemy fire to rescue the downed helicopter's crew.[3] Their awards were subsequently downgraded to Distinguished Flying Crosses.

AIRCRAFT PAINTING AND MARKINGS

The unit arrived in Vietnam with its UH-1A's painted in Gloss Olive Drab (FS. 14087) carrying the Army's standard high visibility markings. It is believed that these UH-1A's carried the unit insignia, the Indian Chief's head described below. The UH-1A's were operated until approximately April 1963.

In March 1963 the unit received its first UH-1B's which were delivered by the USNS Brostrom. The unit was known to have used the following UH-1B aircraft, serial numbers 60-3554, 62-1893 and 62-1897 (and probably 62-1894, 62-1895 and 62-1896).[4] Most of these UH-1B's retained the standard, gloss olive drab paint scheme with high visibility markings.

Unfortunately no photographs of the unit's UH-1A helicopters have been found or identified. Colonel Robert L. Hoffman, USA Ret. who was the second commander of the 45th Transportation Battalion, photographed one of the unit's UH-1B's, s/n 60-3554, while on a medevac mission for a wounded Vietnamese. The distinct unit insignia, an Indian Chief's head, in red, black and white, which originated with the UH-1 Flight Section was carried on the base of the tail fin.

UH-1B, s/n 60-3554, picks up a wounded Vietnamese. Col. Hoffman gave the unit much credit in its conduct of the mission to support the MAAG advisors in all Corps areas,
Photograph courtesy of Col. Robert L. Hoffman, USA Ret.

Right: **Captain Sam Conley, Jr., Transportation Corps, Commander, 611th Transportation Company (DS) in early 1963. Captain Conley is wearing the first pattern utility uniform or "fatigues" with straight pocket flaps, black web belt with brass buckle and black leather service boots. He wears the first design 611th unit patch on the right front pocket of his uniform and the unit's 'signature' black hat with 611th Trans Co (DS) embroidered in gold thread in an arc over his embroidered title Commanders also in gold thread; his rank insignia, silver metal Captain's bars, is attached to the cap front. He has full color sewn on cloth collar rank insignia in white, the Transportation Corps emblem in yellow, cloth Army aviator wings in white, and US Army and name tapes.**

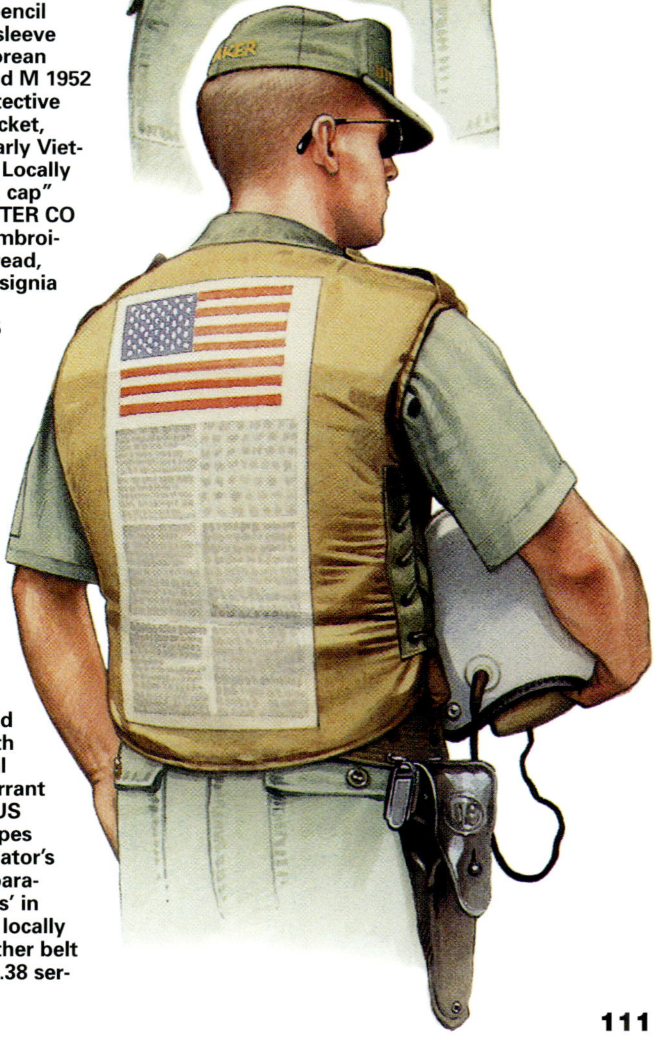

Below Right: **Captain Roger Baker, Artillery, aviator, Utility Tactical Transport Helicopter Company wearing the utility uniform "fatigues" with short sleeves (note the pencil pocket on the left sleeve characteristic of Korean made uniforms) and M 1952 fragmentation protective body armor, flak jacket, adorned with an early Vietnam "Blood Chit". Locally procured "baseball cap" with UTT HELICOPTER CO (variations exist) embroidered in yellow thread, white cloth rank insignia and Army aviator wings, white APH5 flying helmet, sunglasses, date Nov/Dec 1962.**

Left: **CWO Charles E. Larkin, aviator, 57th Transportation Company (Lt Hel), in the Army's gray K2B flight suit with the second design unit patch and standard black leather service boots in mid 1962. Sewn on cloth insignia were in full color, rank and Warrant Officer's emblem, US Army and name tapes and cloth Army aviator's wings and senior parachutist 'jump wings' in white. He carries a locally procured black leather belt and holster for his .38 service revolver.**

APPENDIX 1

Explanation of U. S. Army Aircraft Designation Systems and Serial Number Systems.

U.S. ARMY AIRCRAFT DESIGNATION SYSTEMS

From 1956 through June 1962 the U.S. Army utilized a two-letter system to designate its aircraft as follows:

Letter Code	Explanation	Example
AC	Aircraft, Cargo	AC-1 Caribou
AO	Aircraft, Observation	AO-1 Mohawk
AU	Aircraft, Utility	
HC	Helicopter, Cargo	HC-1B Chinook
HO	Helicopter, Observation	HO-6 (Hughes)
HU	Helicopter, Utility	HU-1A Iroquois 'Huey'

In general usage and practice, helicopters, such as the Shawnee, H-21, or the Sioux, H-13, were most often referred to by the single letter 'H' and the Bird Dog, L-19, and Beaver, L-20, still carried the old U.S. Army Air Force 'L' designation. Army aviators would further refine this terminology by referring to their aircraft using a phonetic equivalent, the H-21 was known as the Hog two-one.

On 6 July 1962 the Department of Defense (DoD) issued DoD Directive 4505.6 Subject: Designation, Redesignation and Naming of Military Aircraft. This directive was issued in order to provide common aircraft nomenclature to be used by all the services; for example the now redesignated UH-1 'Huey' series would be so designated in the Army, the Air Force and the Navy/Marines.

New Letter Code	Explanation	Example
CH	Cargo Helicopter	CH-21C Shawnee
UH	Utility Helicopter	UH-1B Iroquois 'Huey'
OH	Observation Helicopter	OH-13E Sioux
CV	Cargo VTOL/STOL	CV-2B Caribou
OV	Observation VTOL/STOL	OV-1A Mohawk

The old, pre July 1962, 'L' designation was replaced by both 'O' and 'U' as in:

Observation O-1D Bird Dog

and,

Utility U-6A Beaver

The above having been said, this present volume, covering the time frame October 1961 to mid 1963, predominantly uses the 'old', pre-July 1962 aircraft designations, for the sake of continuity in the text. All future volumes will refer to Army aircraft by the post-July 1962 designations.

U.S. ARMY AIRCRAFT SERIAL NUMBER SYSTEMS

Often referred to as the aircraft's 'Radio Call Numbers' in the Army's Technical Bulletins, these numbers were derived from the serial number assigned to the aircraft by the manufacturer. This, in essence, was the identification number, which distinguished that particular aircraft from all similar aircraft in the Army's inventory. To this day, many veteran Army aviators can refer to their aircraft of twenty, thirty or almost forty years ago by the 'last three' of the serial number.

In 1956, the minimum number of digits in the serial number displayed on the tail or fin of an Army aircraft was set at five. An aircraft would carry five, or more, digits composed of the last digit of the fiscal year the aircraft was procured and the last four (or more) digits of the aircraft's identification (serial) number:

For example, an 81st Transportation Company (Lt Hel) H-21C with the serial number:

U.S. ARMY
115896

was procured in fiscal year 195<u>1</u> and was serial numbered <u>15896</u> (from a batch of 33 H-21C's procured in 1951 from Vertol serial numbered 15881 through 15913).

In 1957, the last change was instituted, when the maximum number of digits in the serial number on the tail or fin was set at five. This standardized using the last five of the serial number, which may or may not have included the last digit of the fiscal year of procurement.

In the following example, an 18th Aviation Company U-1A Otter with the serial number:

U.S. ARMY
81713

was procured in fiscal year 195<u>8</u> and was serial numbered <u>1713</u> (from a batch of 40 U-1A's procured in 1958 from De Havilland of Canada serial numbered 1681 through 1720).

With minor exceptions, the serial number system introduced in 1957 was used on Army aircraft throughout the years in Vietnam.

APPENDIX 2

"ON TOP OF OLD AP BAC"

A parody, sung to the tune of "On Top of Old Smokey", written by
CWO Pappy Clayton, 93rd Transportation Company (Lt Hel),
after the battle of Ap Bac on 2 January 1963.
[Verses courtesy of Frank Baldwin, 93rd Trans. Co. (Lt Hel)]

1. We were called into Tan Hiep
   On January Two
   We would never have gone there
   If we'd only knew

2. We were supporting the ARVN's
   A group without guts
   Attacking a village
   Of straw covered huts.

3. A ten copter mission
   A hundred troop load
   Three lifts are now over
   A fourth on the road.

4. The VC start shooting
   They fire a big blast
   We off load the ARVN's
   They sit on their ass.

5. One copter is crippled
   Another sits down
   Attempting a rescue
   Now there are two on the ground.

6. A "Huey" returns now
   To give them some aid
   The VC are so accurate
   They shoot off a blade

7. Four pilots are wounded
   Two crewmen are dead
   When its all over
   A good day for the red.

8. An armored Battalion
   Just stayed in a trance
   One Captain died trying
   To make them advance.

9. The paratroops landed
   A magnificent sight
   There was hand to hand combat
   But no VC in sight.

10. When the news was reported
    The ARVN's had won
    The VC are laughing
    Over their captured guns.

11. The newsmen are trying
    To get some true words
    For the news was reported
    Is all for the birds.

12. All pilots take warning
    When tree lines are near
    Lets land those damn copters
    One mile to the rear.

APPENDIX 3

A ribald compilation of songs gathered by the 57th Transportation Company (Lt Hel) but could just as easily have been sung in any gathering of two or three officer and WO pilots in the country in mid 1963. [Lyrics courtesy of David Prewitt, 57th Trans. Co. (Lt Hel)]

GIVE ME OPERATIONS

Give me operations way out some lonely atoll,
For I'm too young to die, I just want to go home.

Don't give me a UH-1A,
Not a nickel would I pay.
With one rocket blast, it will shoot off your ass,
Don't give me a UH-1A —
OOOOOhh

Don't give me a UH-1B,
Its a hell of a ship don't you see.
Its Emerson kit, it will not shoot shit,
Don't give me a UH-1B —
OOOOOhh

Don't give me an OV-1,
Special weapons like radar and guns.
She's fast I don't care, she blows up in mid air,
Don't give me an OV-1 —
OOOOOhh

Don't give me a CV-2,
It's a hell of a ship thru and thru.
With its rate of descent, its gear will be bent,
Before its first PE is due —
OOOOOhh

Just give me a shaggy Shawnee,
It's a hell of a ship don't you see.
With skill and finesse it will fly with the best,
Oh give me a shaggy Shawnee —
OOOOOhh

BLESS THEM ALL

Bless them all, Bless them all,
The long and the short and the tall.
Bless old man Piasecki for building this plane,
But I know a guy who is cursing in vain.
He took off with his balls to the wall,
Lost manifold pressure and all.
The Son of a Bitch - It ran out of Pitch,
Cheer up my lads, bless them all.

Bless them all, Bless them all,
The needle, the airspeed, the ball.
Bless all those instructors who taught us to fly,
Sent us to solo and left us to die.
If ever your blade tip should stall
Well you're in for one hell of a fall.
Buy lilies or Violets for dead chopper pilots,
Cheer up my lads, Bless them all.

Bless them all, Bless them all
The long and the short and the tall.
Bless all the Majors and their bloody sons,
Bless all the Colonels, the fat headed ones.
I'm saying goodbye to them all,
The long and the short and the tall.
Here's to you and lots of others,
You can shove it up brothers.
I'm going home in the fall.

HALLELUIA

Chorus:

Oh Halleluia, Oh Halleluia,
Throw a nickel on the grass, save a chopper pilots ass.
Oh Halleluia, Oh Halleluia,
Throw a nickel on the grass and you'll be saved

I was cruising down the Mekong doing sixty, and five per,
When a call came from Major - O won't you save me Sir.
Got three holes in my rotor blades - my tanks ain't got no gas,
Mayday, Mayday, Mayday, I got V.C. on my ass! !

I put the pitch stick on the floor, to me it looked all right,
The airspeed read one hundred, I really racked it tight.
The airframe gave a shudder, the engine gave a wheeze,
Mayday, Mayday, Mayday, Crash instructions please! !

I just fell through on final and the pilot gave a shout,
He was yelling at the Crew Chief - Oh get those Bastards out!
I pulled in pitch and power - the RPM did fall -
Mayday, Mayday, Mayday, they did hear me call! !

They sent me down to Tan Hiep, they said will be a breeze,
But by the time I got there, there were V.C. in the trees.
My engine coughed and sputtered, it was too shot up to fly,
Mayday, Mayday, Mayday, I'm too young to die! !

I climbed out from the Shawnee, loaded down with guns,
Struggling through the water, I thought my time had come.
The other chopper landed, it was a pretty sight,
Baby, Baby, Baby, We'll see Saigon tonight! !

NOTES

SECTION I NOTES

[1]Document of 8th Military History Detachment, 1st Aviation Brigade, dated May 1968
And extract of Command Airplane Company Unit History, 1970

[2]The Pentagon Papers, Volume Two, Beacon Press - Document 104, excerpts from General Taylor's report of 3 November 1961, p. 653

[3]The Pentagon Papers, Volume Two, Beacon Press, p. 112

SECTION II NOTES

[1]The New Face of War by Malcolm W. Browne - The Bobbs-Merrill Company, Inc. 1965 p. 53

[2]Letter from Colonel Emmett F. Knight, USA Ret. to author September 1998

[3]A Page in History by Major Milton P. Cherne - U.S. Army Aviation Digest May 1963

[4]A Page in History by Major Milton P. Cherne - U.S. Army Aviation Digest May 1963
And letter from Colonel Emmett F. Knight, USA Ret. to author September 1998

[5]121st Aviation Company (Air Mobile Light) Unit History 1 Jan - 31 Dec 1965

[6]Logistical Support of Airmobile Operations, Republic of Vietnam 1961-1971 by Major Bobby D. Harber - U.S. Army Aviation System Command, March 1971 p. 9

[7]Command and Control, Vietnam Studies Series by Major General George S. Eckhardt 1974 p. 33

[8]18th Aviation Company (UAC) Unit History 1 Jan - 31 Dec 1968

[9]Helicopters Over South Vietnam by Dickey Chapelle - National Geographic November 1962
And The Marines in Vietnam 1954 -1973 An Anthology and Annotated Bibliography - History and Museums Division Hqs., USMC 1974 p. 19

[10]Vietnam Inside Story of the Guerilla War by Wilfred G. Burchett - International Publishers Co. 1965 p. 20

[11]Vietnam Diary by Richard Tregaskis - Holt, Rinehart and Winston 1963 p. 108

[12]45th Trans Bn Named Outstanding Aviation Unit - Army Aviation Magazine, Nov 1962

[13]Vietnam Reports by Lt. Col. Kenneth Mertel - Army Aviation Magazine February 1963

[14]Interview with Major George Tracy, USA Ret. formerly with the 45th Transportation Battalion (Trans Acft), August 1971

[15]The Marines in Vietnam 1954-1973 An Anthology and Annotated Bibliography - History and Museums Division, Hqs., USMC 1974 p. 18

[16]57th Transportation Company (Lt Hel) Unit History 1 Jan - 25 June 1963

[17]Article in Vertiflite by Emmett F. Knight - Boeing Vertol Company Magazine May-June 1987 and reprinted in 57th Trans. Co. Scrapbook compiled by Col. Emmett F. Knight, USA Ret. p. 51

[18]57th Transportation Company (Lt Hel) Unit History 1 Jan - 25 June 1963
And General Order 83 dated 27 December 1968 Lists: Aerial Door Gunner Detachments (Provisional) and Door Gunner Platoons (Provisional)

[19]Letter from Colonel Kenneth D. Mertel, USA Ret. to author September 1998

[20]Operational Report -Lessons Learned (ORLL) - 1st Aviation Brigade interview on 18 January 1971 of Lieutenant Colonel Jack A. Walker, Commanding Officer of the 10th Combat Aviation Battalion.

[21]Letter from Colonel Kenneth D. Mertel, USA Ret. to author September 1998

[22]Interview with John Cross formerly with the 23rd SWAD - February 1970

[23]Vietnam Choppers, Helicopters in Battle 1950 - 1975 by Simon Dunstan - Osprey Publishing Ltd. 1988 pp. 27 - 28
And A Bright Shining Lie, John Paul Vann and America in Vietnam by Neil Sheehan Random House 1988 p. 216

[24]America's Fighting Sky Cavalry by CWO Henry R. Beau - Argosy, June 1963 p. 120

[25]UTT Co. Unit History - Rocket, October 1962 - January 1964

[26]A Bright Shining Lie, John Paul Vann and America in Vietnam by Neil Sheehan Random House 1988 p. 236

[27]America's Fighting Sky Cavalry by CWO Henry R. Beau - Argosy, June 1963 p. 120

[28]Vietnam Diary by Richard Tregaskis - Holt, Rinehart and Winston 1963 p. 376

[29]South Vietnam: On Top of Old Ap Bac - article in Newsweek Magazine - April 1963

[30]'Victory at Ap Bac' from a set of two NLF stamps issued in late 1963 – Michel Catalog #4-5.

[31]Ap Bac - Major Battles in South Vietnam 1963 - 1964 - Foreign Languages Publishing House, Hanoi 1965

[32]Unpublished narrative of Col. Bernard B. Quedens, USA Ret. formerly of the 120th Aviation Company.

[33]Airmobility 1961 - 1971, Vietnam Studies by Lieutenant General John J. Tolson - p. 30

[34]UTT Co. Unit History - Rocket, October 1962 - January 1964

[35]57th Transportation Company (Lt Hel) Unit History 1 Jan - 25 June 1963

[36]145th Combat Aviation Battalion Unit History 1 July 1965 - 30 November 1966

SECTION III NOTES

[1]The Advisory Years To 1965, The United States Air Force in Southeast Asia by Robert F. Futrell, Office of Air Force History, USAF 1981 p. 81

[2]Marines and Helicopters 1962 - 1973 by LTC William R. Fails, History and Museums Division, USMC 1978 p. 32

SECTION IV NOTES

[1]Letter from Harry L. Ammlung, Director, Coating and Chemical Laboratory, U.S. Army Mobility Equipment Research & Development Center to author dated 27 November 1972
And Standard Color Card for Olive Drab No. X34087 (Lusterless), made Standard 23 December 1959

[2]The Pentagon Papers, Volume Two, Beacon Press, p. 13 (Saigon Message 488)
And Passing the Torch, the Vietnam Experience - Boston Publishing 1981 p. 192

[3]Letter from LTC. James E. Gray, Jr., USA Ret. to the author on 31 March 1999

[4]First red fuselage stripe markings observed on H-21, aircraft 'U', of the 93rd Transportation Company (Lt Hel) photographed by Dickey Chapelle in mid 1962 and published in National Geographic Magazine in November 1962.

[5]Letter from CW4 Frank Baldwin, USA Ret. formerly of the 93rd Transportation Company (Lt Hel) to author October 1973

[6]Army Aviation, Special Report - Flying, Vol. 70, No. 5, May 1962 photograph p. 22

SECTION V NOTES

45TH TRANSPORTATION BATTALION (TRANSPORTATION AIRCRAFT)

[1]45th Trans Bn Named Outstanding Aviation Unit - Army Aviation Magazine Nov 1962

[2]Vietnam Reports by Lt. Col. K. D. Mertel - Army Aviation Magazine Aug-Sep 1963

[3]Letter from CW4 Frank Baldwin, USA Ret., formerly of the 93rd Transportation Company (Lt Hel) to author September 1973

[4]Personal knowledge of CW4 Frank Baldwin, USA Ret., formerly of the 93rd Transportation Company (Lt Hel).

18TH AIRFIELD OPERATING DETACHMENT

[1]57th Transportation Company (Lt Hel) Unit History 1 Jan - 25 June 1963

57TH TRANSPORTATION COMPANY (LIGHT HELICOPTER)

[1]Letter from Col. Emmett F. Knight, USA Ret. to author September 1998

[2]Scrapbook of 57th Trans Co., 98th Trans Det. and CH-21C by Col. Emmett F. Knight, USA Ret. Boeing Vertol Company, 1987 p.73

[3]'Bananas' Manned by Fighting Bunch by Malcolm W. Browne - Article in Pacific Stars and Stripes, 31 October 1962
And The New Face of War by Malcolm W. Browne - The Bobbs-Merrill Company, Inc. 1965. p. 57

[4]Scrapbook of 57th Trans Co., 98th Trans Det. and CH-21C by Col. Emmett F. Knight, USA Ret. Boeing Vertol Company, 1987 p.80

[5]Vietnam Diary by Richard Tregaskis - Holt, Rinehart and Winston 1963 p. 166

[6]Scrapbook of 57th Trans Co., 98th Trans Det. and CH-21C by Col. Emmett F. Knight, USA Ret. Boeing Vertol Company, 1987 p. 45

[7]57th Transportation Company (Lt Hel) Unit History 1 Jan 63 - 25 Jun 63.

[8]57th Transportation Company (Lt Hel) Unit History 1 Jan 63 - 25 Jun 63.

[9]21 SKID-DOO - Pictorial (Rear Cover) - United States Aviation Digest, April 1964
And Scrapbook of 57th Trans Co., 98th Trans Det. and CH-21C by Col. Emmett F. Knight, USA Ret. Boeing Vertol Company, 1987 p. 41
And List of "FIRST" for the 57th Transportation Company (Lt Hel) (H-21) unpublished document dated 10 August 1962

[10]Letter from Col. William C. Rudd, USA Ret. to author 23 January 1999

[11]Letter from Col. Emmett F. Knight, USA Ret. to author September 1998

[12]57th Transportation Company (Lt Hel) Unit History 1 Jan 63 - 25 Jun 63.

[13]Report by Colonel John L. Klingenhagen - Army Aviation Magazine Jan-Feb 1964

[14]Letter from Col. Emmett F. Knight, USA Ret. to author September 1998

[15]Army Aviation, Special Report - Flying Vol. 70, No. 5, May 1962

8TH TRANSPORTATION COMPANY (LIGHT HELICOPTER)

[1]Vietnam Report by Lt. Col. Kenneth D. Mertel - Army Aviation Magazine June - July 1963

[2]America's Fighting Sky Cavalry by Henry R. Beau - Argosy Magazine June 1963 p. 122

[3]America's Fighting Sky Cavalry by Henry R. Beau - Argosy Magazine June 1963

[4]Telephone interview with LTC. Robert F. Corneil, USA Ret. 19 April 1999

[5]Vietnam Diary by Richard Tregaskis - Holt, Rinehart and Winston 1963 p. 369

93RD TRANSPORTATION COMPANY (LIGHT HELICOPTER)

[1]Letter from LTC. James E. Gray, Jr., USA Ret. to author 31 March 1999

[2]Vietnam Diary by Richard Tregaskis -Holt, Rinehart and Winston 1963 p. 157

[3]Telephone interview with Col. Sam Conley, Jr., USA Ret. October 1998

[4]Vietnam Diary by Richard Tregaskis - Holt, Rinehart and Winston 1963 p. 157

[5]Telephone interview with Col. Sam Conley, Jr., USA Ret. 20 November 1998

[6]Army Aviation Progress in Vietnam by Lt. Col. Kenneth D. Mertel - AA Magazine July - August, 1963

[7]Vietnam Diary by Richard Tregaskis - Holt, Rinehart and Winston 1963 p. 15

[8]Letter from LTC Leon W. Curry, USA Ret. to author 2 March 1999

[9]Vietnam Diary by Richard Tregaskis - Holt, Rinehart and Winston 1963 p. 308

[10]According to Col. Sam Conley, Jr. USA Ret. and CWO John Leonard, USA Ret., formerly of the 611th Transportation Company (DS), both of whom assisted in the recovery of aircraft following the battle of Ap Bac, there were a total of six H-21's and one UH-1B downed by enemy action on 2 January 1963. Letter from Col. Sam Conley, Jr. USA Ret. to the author dated 15 December 1998.

[11]Vietnam Front Pages, Pacific Stars and Stripes edited by Hal Drake - Bonanza Books 1986. Dust cover color photograph of an H-21, in standard high visibility markings, carries red flight platoon nose marking, with white trim, and red fairings top and bottom on the vertical stabilizer. What appears to be a white aircraft identification letter is carried on the tail. There is also a personal marking painted behind the co-pilot's window.
And Vietnam – A Pictorial Review From the Pages of LIFE Magazine - 1965. On page 38, in a photograph by Larry Burrows, a 93rd Transportation Company (Lt Hel) H-21, serial number unknown, aircraft letter white 'P' carries red flight platoon markings, with white trim, on the nose triangle formed by the canopy framing.

[12]Helicopter War in South Vietnam by Dickey Chapelle - National Geographic November 1962. This article, written in mid 1962, illustrates aircraft 'U', s/n unknown, of the 93rd Trans. Co. on the ground at an outpost near the Laotian border. Aircraft carries a red conspicuity stripe on the fuselage top, page 743.

33RD TRANSPORTATION COMPANY (LIGHT HELICOPTER)

[1]This pocket patch was donated by Col. Thomas E. Connell, USA Ret. to the US Army Transportation Museum. It is unusual in that it depicts a Huey helicopter and not the Shawnee helicopter usually displayed on the Transportation Detachment (CHFM) pocket patches in use in late 1961 to mid 1963 in Vietnam.

[2]Vietnam Report by Lt. Col. Kenneth D. Mertel - Army Aviation Magazine February 1963

[3]Letter from Col. Sam Conley, Jr., USA Ret. to author September 1998

[4]Letter from LTC. Joseph E. Henderson, USA Ret. to author February 1983

[5]Letter from LTC. Joseph E. Henderson, USA Ret. to author February 1983

81ST TRANSPORTATION COMPANY (LIGHT HELICOPTER)

[1]Vietnam Diary by Richard Tregaskis - Holt, Rinehart and Winston 1963 p. 313

[2]Telephone interview with Col. Sam Conley, Jr., USA Ret. on 20 November 1998.

[3]Vietnam Diary by Richard Tregaskis - Holt, Rinehart and Winston 1963 pp. 353 - 356

[4]From copy of a letter to James D. Sprinkle from Col. Paul F. Anderson.

[5]A Pictorial History of the Vietnam War by Richard F. Newcomb - Doubleday & Company, Inc. 1987. Two photographs of General Stilwell's dedication of Camp Holloway on 4 July 1963 are shown on page 69.

UTILITY TACTICAL TRANSPORT HELICOPTER COMPANY

[1]Letter to James D. Sprinkle from the Editor of U.S. Army Aviation Digest, dated 20 September 1973

[2]Scrapbook 57th Trans. Co. by Col. E.F. Knight, USA Ret. - Boeing Vertol Co. 1987 p. 80

[3]The Armed Helicopter Story Part III, Armed Helicopters Around the World by LTC Charles O. Griminger - U.S. Army Aviation Digest - Vol. 17, No. 9 September 1971 p. 22

[4]U.S. Army Aviation Digest - Vol. 20, No. 3 March 1974 p. 29

[5]Views from Readers, U.S. Army Aviation Digest - Vol. 20, No. 7 July 1974 p. 1

[6]Scrapbook 57th Trans. Co. by Col. Emmett F. Knight, USA Ret. - Boeing Vertol Co. 1987 p. 74

[7]A Bright Shining Lie, John Paul Vann and America in Vietnam by Neil Sheehan - Random House 1988 p. 219

[8]Vietnam Diary by Richard Tregaskis - Holt, Rinehart and Winston 1963 p. 152

[9]Airmobility 1961 -1971 by Lt. Gen. John J. Tolson - Vietnam Studies p. 30

[10]Vietnam Report by Lt. Col. Kenneth D. Mertel - Army Aviation Magazine March 1963

[11]Unnamed Army aviator quoted in 'Viet Copter War - the Price is High' Pacific Stars and Stripes 14 January 1963 p. 6

[12]Army Gunships in Vietnam by Bob Chenoweth - Arms and Armour Press, 1987 Warbirds Illustrated No. 47; photograph of UH-1A "Ole Hotbox" on p. 13

[13]Airmobility 1961 - 1971 by Lt. Gen. John J. Tolson - Vietnam Studies p. 35

[14]Army Gunships in Vietnam by Bob Chenoweth - Warbirds Illustrated No. 47 p. 17

23RD SPECIAL WARFARE AVIATION DETACHMENT

[1]Airmobility 1961-1971 by Lt. Gen. John J. Tolson - Vietnam Studies p. 41

[2]The Mohawk in Vietnam - Army Aviation Magazine 24 February 1967 p. 18

[3]Vietnam Report by Col. John L. Klingenhagen - Army Aviation Magazine Jan-Feb 1964

[4]Airmobility 1961-1971 by Lt. Gen. John J.Tolson - Vietnam Studies p. 42

[5]Letter from Col. Sam Conley, Jr. USA Ret. to author 15 December 1998

[6]Airmobility 1961-1971 by Lt. Gen. John J. Tolson - Vietnam Studies p. 43

[7]Airmobility 1961-1971 by Lt. Gen. John J. Tolson - Vietnam Studies p. 42

[8]Interview with John Cross, formerly with 23rd SWAD - February 1970

611TH TRANSPORTATION COMPANY (DIRECT SUPPORT)

[1]Letter from Col. Sam Conley, Jr. USA Ret. to author 15 December 1998

[2]Vietnam Report by Lt. Col. Kenneth D. Mertel - Army Aviation Magazine April 1963

[3]General Orders No. 177, dated 26 April 1963 - Headquarters US Army, Pacific

[4]Vietnam Report by Lt. Col. Kenneth D. Mertel - Army Aviation Magazine April 1963

[5]Telephone interview with Col. Sam Conley, Jr., USA Ret. October 1998

[6]Letter from Col. Sam Conley, Jr., USA Ret. to author October 1998

[7]Telephone interview with Col. Sam Conley Jr., USA Ret. October 1998

[8]57th Transportation Company (Lt Hel) Unit History 1 Jan - 25 June 1963

[9]Vietnam Report by Lt. Col. Kenneth D. Mertel - Army Aviation Magazine April 1963

[10]Letter from LTC. Charles P. Callaway, USA Ret. to author February 1983

[11]Support Command, Vietnam - Army Aviation Magazine January 1965

AVIATION SECTION, 3RD RADIO RESEARCH UNIT

[1]Copy of General Order 48, HQ USASA dated 20 May 1966 in the National Archives

[2]From Internet Posting on 3rd Radio Research Unit by Dennis Buley, 1998

[3]An American In The War Before '64 by J. T. Chapin - Vietnam October 1995

[4]From Internet Posting on 3rd Radio Research Unit by Dennis Buley, 1998

[5]Letter from Dennis Buley to author dated 27 November 1998
And Army Neptunes Over SE Asia by Wayne Mutza - Air Enthusiast No. Twenty-nine. Article mentions ARDF systems installed in three Beavers of the Vietnam Flight Detachment. In March 1962 the MACV-MAAG Flight Detachment may have supplied the three U-6A's.

[6]From Internet Posting on 3rd Radio Research Unit by Dennis Buley, 1998 The only U-8F known to have been operated by Radio Research units in Vietnam was aircraft s/n 63-13637, which may have been the same U-8F received and operated by the Aviation Section, 3rd RRU in 1963.

[7]Army Neptunes Over SE Asia by Wayne Mutza - Air Enthusiast No. Twenty-nine.

[8]An American In The War Before '64 by J. T. Chapin - Vietnam October 1995

[9]Letter and information received from INSCOM Command Historian dated 7 December 1998.

[10]Telephone interview of Dennis Buley on 18 November 1998.

1ST AVIATION COMPANY (CV-2)

[1]Vietnam Report by Lt. Col. Kenneth D. Mertel - Army Aviation Magazine May 1963

[2]Vietnam Report by Col. John L. Klingenhagen - Army Aviation Magazine Jan-Feb 1964

[3]Telephone interview with Mel Brooke, former member of the 1st Aviation Company (CV-2) on 20 March 1983.

[4]Vietnam Report by Lt. Col. Kenneth D. Mertel - Army Aviation Magazine April 1963

[5]Vietnam Diary by Richard Tregaskis - Holt, Rinehart and Winston 1963 p. 191

[6]Vietnam Report by Lt. Col. Kenneth D. Mertel - Army Aviation Magazine April 1963

[7]Vietnam Report by Lt. Col. Kenneth D. Mertel - Army Aviation Magazine April 1963

[8]C-7 Caribou in Action by Wayne Mutza - Squadron Signal No. 132 p.13

[9]Vietnam Report by Col. John L. Klingenhagen - Army Aviation Magazine Jan-Feb 1964

MACV-MAAG FLIGHT DETACHMENT

[1]The Laos Story by Major Leland F. Wilhelm - Army Aviation Magazine

[2]Vietnam Report by Lt. Col. Kenneth D. Mertel - Army Aviation Magazine April 1963

[3]Telephone interview of John Leonard, CWO USA Ret. formerly with the 611th Transportation Company (DS) in September 1998.

[4]Command Airplane Company Unit History - Extract dated 1970.

[5]Painting and Marking of Army Aircraft - TB 746-93-2 January 1971 Change 1 dated 21 October 1971, Chapter 7-38 American Flag Decals. p. 7-12

57TH MEDICAL DETACHMENT (HELICOPTER AMBULANCE) (UH-1A)

[1]Dust Off by John L. Cook - The Illustrated History of the Vietnam War p. 153

[2]Dust Off by John L. Cook - The Illustrated History of the Vietnam War p. 48

[3]Dust Off by John L. Cook - The Illustrated History of the Vietnam War p. 24

[4]Vietnam Report by Lt. Col. Kenneth D. Mertel - Army Aviation Magazine March 1963

[5]Dust Off by John L. Cook - The Illustrated History of the Vietnam War p. 41

[6]Dust Off by John L. Cook - The Illustrated History of the Vietnam War p. 32

[7]Dust Off by John L. Cook - The Illustrated History of the Vietnam War p. 49

[8]Dust Off by John L. Cook - The Illustrated History of the Vietnam War p. 25 UH-1A, s/n believed to be 58-2082, of the 57th Medical Detachment (Hel Amb).

A FIGHT, 1ST PLATOON, 19TH TRANSPORTATION COMPANY (H-37)

[1]Letter from LTC. Charles P. Callaway, USA Ret. to author February 1983

[2]Letter from LTC. Charles P. Callaway, USA Ret. to author 12 January 1999

[3]Letter from LTC. Charles P. Callaway, USA Ret. to author February 1983

[4]Letter from LTC. Charles P. Callaway, USA Ret. to author February 1983

[5]Letter from LTC. Charles P. Callaway, USA Ret. to author 12 January 1999

UH-1 FLIGHT SECTION, 45TH TRANSPORTATION BATTALION

[1]Letter from Col. Robert L. Hoffman, USA Ret. to author 22 April 1999

[2]US Army General Order 83, dated 27 December 1968, Vietnam Advisory Campaign.

[a]Combat Zone: Territorial limits of Vietnam and adjacent waters.

[b]Time limitations: 15 March 1962 - 7 March 1965

[c]Unit participation: GO 83 lists a 1st Aviation Platoon and a 4th Aviation Detachment which can not presently be matched to any known or documented aviation unit which served in Vietnam during that time frame.

[3]Letter from LTC. James E. Gray, Jr., USA Ret. to author 31 March 1999
And Vietnam Diary by Richard Tregaskis - Holt, Rinehart and Winston p. 47

GLOSSARY of commonly used acronyms

ACTIV — Army Concept Team in Vietnam

AOD — Airfield Operating Detachment

AR — Army Regulation

ARDF — Airborne Radio Direction Finding

ARPA — Advanced Research Project Agency

ARVN — Army Republic of Vietnam

ASA — Army Security Agency (also USASA)

ASTA — Aerial Surveillance and Target Acquisition

ATC — Air Traffic Control

Bn — Battalion

C&C — Command and Control

CHFM — Cargo Helicopter Field Maintenance

CIA — Central Intelligence Agency

CO — Commanding Officer

CONUS — Continental United States

COSVN — Central Office for South Vietnam

CWO — Chief Warrant Officer

DA — Department of the Army

DFC — Distinguished Flying Cross

DFO — Direction Finding Operator

DMZ — Demilitarized Zone

DoD — Department of Defense

DS — Direct Support

EM — Enlisted Man or Men

FCC — Flight Coordination Center

FFS — Flight Following Station

FOC — Flight Operations Center

FS — Federal Standard

FW-LT — Fixed Wing - Light Transport

FW-TT — Fixed Wing - Tactical Transport

GCA — Ground Control Approach Radar System

GO — General Order

GS — General Support

HMM — Helicopter Marine Medium

HP — Hot Pilot

IG — Inspector General

INSCOM — Intelligence and Security Command

LPH — Amphibious Assault Ship (U.S. Navy)

LZ — Landing Zone

MAAG — Military Assistance Advisory Group

MACV — Military Assistance Command, Vietnam

NATO — North Atlantic Treaty Organization

NBS — National Bureau of Standards

NCO — Noncommissioned Officer

NDB — Non-Directional Beacon

NLF — National Liberation Front

NSA — National Security Agency

OPLAN — Operational Plan

ORLL — Operational Report - Lessons Learned

PE — Periodic Inspection

PFC — Private First Class

PSP — Perforated Steel Planking

RPM — Revolutions Per Minute

R&R — Rest and Recreation

RRU — Radio Research Unit

RVN — Republic of Vietnam

SEATO — Southeast Asia Treaty Organization

SFC — Sergeant First Class

s/n — Serial Number

SOI — Signal Operations Instructions

SOP — Standard Operating Procedure

STOL — Short Take-Off & Landing

SWAD — Special Warfare Aviation Detachment

TB — Technical Bulletin

TDA — Table of Distribution and Allowances

TDY — Temporary Duty (Assignment)

TOE — Table of Organization and Equipment

TWR — (Control) Tower

U.S. — United States

USA — United States Army

USAF — United States Air Force

USARV — U.S. Army, Vietnam

USARYIS — U.S. Army, Ryukyu Islands

USASA — United States Army Security Agency

USASG-V — U.S. Army Support Group - Vietnam

USMC — United States Marine Corps

USNS — United States Naval Ship (Merchant Marine operated)

USS — United States Ship

UTT — Utility Tactical Transport Helicopter Company

VC — Viet Cong

VHP — Very Hot Pilot

VIP — Very Important Person

VNAF — Vietnamese Air Force

VTOL — Vertical Take-Off & Landing

WO — Warrant Officer

XO — Executive Officer

BIBLIOGRAPHY • Referencing Army Aviation in Vietnam 1961-Mid 1963

BOOKS

Airmobility 1961-1971 by Lieutenant General John J. Tolson - Vietnam Studies series published by the Department of the Army 1973

Although filled with much useful information on the early years of Army aviation in Vietnam, this first Army study of airmobility, and the war, suffers from some inaccuracies.

Ap Bac Major Battles in South Vietnam 1963-1964 by Le Hong Linh, Vuong Thanh Dien and Nguyen Q.S. - Foreign Languages Publishing House, Hanoi 1965

Communist propaganda put out by North Vietnam, however has the best diagram of the overall battlefield of Ap Bac to date.

Passing the Torch, The Vietnam Experience - Boston Publishing Company 1981

Worthwhile series with balanced text, photographs and artwork.

Raising the Stakes, The Vietnam Experience - Boston Publishing Company 1982

As above, but with a few photographs of early Army aviation units in this volume.

Thunder From Above, The Vietnam Experience - Boston Publishing Company 1984

Photograph on page 15 shows rebellious pilot bombing Presidential palace on 26 February 1962, and unusual view of UTT UH-1B, with H-21's in background, downed at Ap Bac.

Command and Control 1950-1969 by Major General George S. Eckhardt - Vietnam Studies series published by the Department of the Army 1974

Provides a good overview of the Army's command and control structure in Vietnam, with names, dates and photographs of key commanders.

The New Face of War by Malcolm W. Browne - The Bobbs-Merrill Company, Inc. 1965

A factual, easy to read report on the war from the Associated Press representative in Vietnam. Some of the material on Army aviation will not be found anywhere else. Browne was trusted by Army aviators to 'get it right.'

Vietnam Diary by Richard Tregaskis - Holt, Rinehart and Winston 1963

World War Two type reporting on Vietnam in 1962, yet provides one-of-a-kind classic coverage of a few Army aviation units .

Army Gunships in Vietnam by Bob Chenoweth - Arms and Armour Press 1987

Excellent coverage of Army gunships mainly UH-1 series, with sizable coverage of the early years

Air War in Vietnam by Philip D. Chinnery - Exeter Books 1987

Good coverage of the air war, some coverage of Army aviation. Well written and researched.

Vietnam the Helicopter War by Philip D. Chinnery - Naval Institute Press 1991

Excellent account, with first person narratives, of the helicopter war. Highly recommended, although the coverage is not just of Army aviation units, but of the other services that flew helicopters too.

Vietnam The War in the Air by Rene J. Francillon - Arch Cape Press 1987

Well respected aviation author, small section on Army aviation covering the entire war

Vietnam Choppers, Helicopters in Battle 1950-1975 by Simon Dunstan - Osprey Publishing Ltd. 1988

Well put together account of helicopters in use by all services. Small, but accurate, section on Army aviation.

The History of the Vietnam War by Douglas Welsh - Galahad Books 1981

One photograph on page 39, top right, pictures the battlefield at Ap Bac from the air, before recovery of the downed 93rd Trans. Co. H-21's and UTT UH-1B; also visible are the tracks of the ARVN M-113's, showing the furthest extent of their attacks.

Vietnam, The Decisive Battles by John Pimlott -Macmillan Publishing Company 1990.

Contains one section covering the events leading up to, and surrounding, the battle of Ap Bac which includes a schematic illustration of the battlefield.

Vietnam, Inside Story of the Guerilla War by Wilfred G. Burchett - International Publishers Co. 1965

An unabashed apologist and propagandist for the communists, Burchett nonetheless manages to convey, at times, the Viet Cong experience of being on the receiving end of a heliborne combat assault. Makes one glad that one was on our side!

Airmobile, The Helicopter War in Vietnam by Jim Mesko - Squadron Signal Publications 1984

Well illustrated, general overview of Army helicopters in Vietnam, minimal coverage of early aviation units.

A Bright Shining Lie, John Paul Vann and America in Vietnam by Neil Sheehan - Random House 1988.

Neil Sheehan covered Ap Bac in Vietnam in 1963 and has written the best overall account of the battle yet published.

Dust Off by John L. Cook - one of the Bantam Book series The Illustrated History of The Vietnam War, 1988

This, and John Cook's later, expanded work, contains excellent coverage, with photographs, of the early days of the 57th Medical Detachment (Helicopter Ambulance).

The French Indochina War 1946-54 by Martin Windrow - Osprey Military Men-At-Arms 322

Compact, illustrated history of the First Indochina War by noted author. Provides the reader with a listing of units, both Viet Minh and French, tactics, a chronology of major operations and color uniform plates.

Logistical Support of Airmobile Operations, Republic of Vietnam 1961-1971 by Major Bobby D. Harber - U.S. Army Aviation System Command, March 1971

The first 'history', from a logistics point of view, of Army aviation in Vietnam, published in 1971 before the end of U.S. involvement. This work remains, to this day, a classic, which has stood up well to the passage of time.

Area Handbook for South Vietnam by co-authors Harvey H. Smith, Donald W. Bernier, Frederica M. Bunge, Frances Chadwick Rintz, Rinn-Sup Shinn and Suzanne Teleki - Department of the Army Pam 550-55 April 1967

Good, one volume reference for history, geography, population, politics, economy, and armed forces of the Republic of South Vietnam, with maps.

Marines and Helicopters 1962-1973 by Lieutenant Colonel William R. Fails, USMC - History and Museums Division, Headquarters, U.S. Marine Corps 1978

Excellent History of USMC Helicopters in Vietnam, contains a great deal of information.

The Marines in Vietnam 1954-1973 An Anthology and Annotated Bibliography -History and Museums Division, Headquarters, USMC 1974

Contains article by Commander of HMM-362 describing unit's tour in Soc Trang.

Images Of A Lengthy War The United States Army in Vietnam by Joel D. Meyerson - Center of Military History, United States Army 1986

One of the official histories published by the Army on Vietnam, this particular volume is devoted to photographs of the war. While very few specifically show Army aviation assets, some of the photographs from the time period 1945 to 1963 are, nonetheless, interesting.

Advice and Support The Early Years The United States Army in Vietnam by Ronald H. Spector - Center of Military History, United States Army 1983

Excellent, early history of the Army's advisory assistance to the French and later the Vietnamese. Background material on the ARVN under President Diem and Viet Cong infiltration of the RVN government and military in 1960.

Tiger In The Barbed Wire, An American in Vietnam 1952-1991 by Howard R. Simpson - Kodansha International 1994.

An excellent, first hand account of the early years of American involvement in Indochina.

The Pentagon Papers The Defense Department History of United States Decisionmaking on Vietnam, Volume One - The Beacon Press

Surrounded by controversy and charges of treason when first made public by Daniel Ellsberg and the New York Times, the Pentagon Papers provide an authoritative and comprehensive source of official documents and decisions on Vietnam and the conduct of the war. Volume I covers the years 1940 -1960

The Pentagon Papers The Defense Department History of United States Decisionmaking on Vietnam, Volume Two - The Beacon Press

As above. Volume II covers the years 1961 to 1967

The Pentagon Papers The Secret History of the Vietnam War by Neil Sheehan, Hedrick Smith, E.W. Kenworthy and Fox Butterfield - As Published by the New York Times July 1971

The material published in the New York Times. While not specific to Army aviation, this paperback contains 64 pages of excellent photographs of key American and Vietnamese players from a historical perspective.

Vietnam Front Pages, Pacific Stars and Stripes edited by Hal Drake - Bonanza Books 1986

Interesting, and informative, copies of the front pages only of select issues of Pacific Stars and Stripes starting with Volume 1, No. 1 in 1945 and ending with Vol. 31, No. 120 in 1975.

U. S. Army Uniforms of the Vietnam War by Shelby Stanton - Stackpole Books 1989

To the usual, thorough standard of this respected author, this useful book covers and illustrates Army aviation flight clothing and equipment as well as everything else that was worn by Army personnel in Vietnam.

PERIODICALS AND NEWSPAPERS

Army Aviation Association of America

- Publication
  Army Aviation Magazine

  Vietnam Report by Lt. Col. Kenneth D. Mertel - January 1963

  Vietnam Report by Lt. Col. Kenneth D. Mertel - February 1963

  Vietnam Report by Lt. Col. Kenneth D. Mertel - March 1963

  Vietnam Report by Lt. Col. Kenneth D. Mertel - April 1963

  Vietnam Report by Lt. Col. Kenneth D. Mertel - May 1963

  Vietnam Report by Lt. Col. Kenneth D. Mertel - June-July 1963

  Vietnam Report by Lt. Col. Kenneth D. Mertel - July-Aug 1963

  Vietnam Report by Lt. Col. Kenneth D. Mertel - Aug-Sept 1963

  Vietnam Report by Lt. Col. Kenneth D. Mertel - Sept-Oct 1963

  Vietnam Report by Lt. Col. Kenneth D. Mertel - Oct-Nov 1963

  Vietnam Report by Lt. Col. Kenneth D. Mertel - Nov-Dec 1963

  Vietnam Report by Col. John L. Klingenhagen - Jan-Feb 1964

With remarkable breadth and scope the "Vietnam Reports" originated, and the majority written, by Lt. Col. Mertel literally provides the reader with a perspective on Army aviation in Vietnam in 1963 that can be found nowhere else. These well written articles open a 'window in time' so that one can read about awards for heroism at Ap Bac, and other acts of individual courage; change of commands; just what did 'milk runs' in a Caribou entail; which unit had which pet; gunners on the UH-1B's and H-21's; Air Traffic Control; Survival Kits, Clothing and much more.

45th Trans Bn Named Outstanding Aviation Unit - November 1962

Article includes the remarks and presentation by the Army Chief of Staff, General Earle G. Wheeler, of the 1962 Hughes Tool Company, Aircraft Division's 'Outstanding Army Aviation Unit' Trophy. Also remarks of Major Milton P. Cherne in accepting for the 45th Trans. Bn.

Support Command, Vietnam - January 1965

A concise overview of the U.S. Army Support Command, Vietnam and Army aviation in Vietnam at the end of 1964, with special emphasis on the 611th Transportation Company (DS) in Vung Tau.

The Laos Story by Major Leland F. Wilhelm (issue unknown)

Army aviation in Laos in the early days of MAAG involvement, an interesting article.

The Mohawk in Vietnam - 24 February 1967

Overview of OV-1 Mohawk units and operations in Vietnam. Especially good information and coverage of the early years, for example the material on the 4th Aerial Surveillance and Target Acquisition (ASTA) Platoon, a very seldom mentioned, and poorly documented, unit.

U.S. Army Aviation Digest

  A Page in History by Major Milton P. Cherne - May 1963

  Your Orders by Major Milton P. Cherne - November 1963

Excellent, award-winning accounts of Army aviators and the first airmobile operations in Vietnam in 1962. Major Cherne, Commanding Officer of the 57th Transportation Company (Lt Hel) in Vietnam, has assured the 57th Transportation Company (Lt Hel) a prominent place in any history of Army aviation in Vietnam.

21 SKID-DOO - April 1964

Rear cover pictorial on the effort to get 'Old 049' of the 57th Trans Co. (Lt Hel) out of a rice paddy and 'towed' to base.

The Armed Helicopter Story Part III: Armed Helicopters Around The World by Lieutenant Colonel Charles O. Griminger - September 1971

Great series of articles covering the background, early experiments, personalities, equipment, machines and weapon systems that led to gunship armament, in all its varieties and permutations.

America's Fighting Sky Cavalry by CWO Henry R. Beau - *Argosy* June 1963

CWO Beau, the first Army aviator wounded in Vietnam, talks about his experiences. This article also provides CWO Carleton J. 'Joe' Nysewander's, flying 'Chalk 6' aircraft 'D' serial number #51-15903, account of his role, and being wounded, at Ap Bac on 2 January 1963.

South Vietnam: On Top of Old Ap Bac - *Newsweek* , April 1963

Short article on battle of Ap Bac, includes abbreviated version of 93rd Trans. Co.'s parody song.

South Vietnam Fights the Red Tide by Peter T. White and W.E. Garrett - *National Geographic* October 1961

Interesting pictorial coverage with map, a National Geographic specialty, of the country at war in 1961.

Helicopters Over South Viet Nam by Dickey Chapelle - *National Geographic* November 1962

War story as only Dickey Chapelle could write and photograph. Some of the most visually beautiful images of helicopters at war ever published.

'Bananas' Manned by Fighting Bunch - *Pacific Stars & Stripes*, 31 October 1962

Article written by Malcolm W. Browne about the 57th Transportation Company (Lt Hel) in which Browne first referred to the men of the 57th as "the deans of the helicopter corps."

Ap Bac - Site of a Bloody Stand for Freedom - *Pacific Stars & Stripes*, 9 January 1963

Report on the 2 January 1963 battle of Ap Bac.

Viet Copter War - the Price is High - *Pacific Stars & Stripes*, 14 January 1963

Various thoughts and comments after the loss of seven American lives on 11 January 1963.

An American In The War Before '64 by J. T. Chapin - *Vietnam* October 1995

An interesting article about the mission, techniques and experiences of an early member of the 3rd RRU who served as a direction finding operator (DFO) in 1961 and 1962.

Army Neptunes Over SE Asia by Wayne Mutza - *Air Enthusiast* No. Twenty-nine

Excellent overview of the Army's 'Radio Research' units in Vietnam, mentioning the 3rd RRU, with concentration on the RP-2E Neptunes of the 1st Radio Research Unit.

Vietnam, Blows, Counterblows, Waits - *Life* Vol. 58, No. 8, 26 February 1965

Cover photograph of NLF stamp commemorating 'victory' at Ap Bac.

Army Aviation, Special Report - *Flying* Vol. 70, No. 5, May 1962

An overview of the state of Army aviation in 1962; lead off photograph is only photograph in article of Army aircraft in Vietnam, showing a 57th Trans. Co. H-21 s/n # 55-4219.

De Havilland U-6A/L-20 "Beaver" in Service with the United States by James Wogstad and Phil Friddell - *Replica In Scale* Vol. 2, No. 2 Winter 1974

Well researched and presented article on the U-6A, with excellent coverage of Beaver floatplanes and minimal Vietnam coverage.

Vietnam, the Helicopter War by Bob Chenoweth - *Scale Aircraft Modelling* Vol. 4, No. 9 June 1982

The first article published in a British scale modeling magazine, to give aircraft modelers a fairly comprehensive account of Army aviation unit markings in Vietnam. While an excellent general overview, there is little coverage of the early years.

PRINTED DOCUMENTS

Cloth Patch of the U.S. Armed Forces, Part VII Cloth Insignia of U.S. Army Aviation Units, Volume 1 Aviation and Transportation Units in Vietnam by Col. James F. Greene, Jr. and Lt. Col. Cecil B. Smyth, Jr. - *American Society of Military Insignia Collectors*

A comprehensive, illustrated patch catalog with some inaccuracies.

Army Aviation Distinctive Insignia - *American Society of Military Insignia Collectors* March 1973

Catalog of Army aviation metal insignia, commonly referred to as 'beercans', mostly from Vietnam. Good coverage with some inaccuracies.

Transportation Distinctive Insignia - *American Society of Military Insignia Collectors*

As above, Transportation unit's metal insignia with good coverage of Vietnam units.

Painting and Marking of Army Aircraft - *Department of the Army Technical Bulletin*

TB 746-93-2 28 June 1967 (Revised as necessary)

Everything anyone ever wanted to know about the official painting and marking of Army aircraft, includes line drawings and maintenance data placement information.

Colors - *Federal Standard* No. 595a 2 January 1968 (Revised as necessary)

The entire Federal Standard colors, on .5 x 1 inch chips, in one manual.

Olive Drab No. X34087 (Lusterless), *Standard Color Card*, Made Standard 23 Dec. 1959.

Color Universal Language and Dictionary of Names by Kenneth L. Kelly - National Bureau of Standards (NBS) Special Publication 440, *U.S. Department of Commerce* December 1976

More than most people will ever want to know about color.

Colors, Exterior, Naval Aircraft: Requirements For - *MIL-C-18263E(Weps)* 11 May 1965
Naval Air Systems Command specifications for the exterior colors to be applied to heavier-than-air and lighter-than-air Naval and Marine Corps aircraft.

The Peoples of Mainland Southeast Asia - *National Geographic Society*, Washington, D.C. March 1971
Authoritative map, included in National Geographic Magazine, with topical war related data, and guide to the ethnic groups, and sub-groups, of Vietnam, Laos and Cambodia.

Republic of Vietnam - Capsule Facts for the Armed Forces - *Department of the Army* Pam 360-759 October 1968
Everything the Government wanted one to know about Vietnam.

The Religions of Vietnam - *MACV Office of Information*, Command Information Pam 11-67 April 1968
Surprisingly comprehensive guide, and background, to religion in Vietnam.

Map of South Vietnam - *U.S. Government Printing Office* Stock Number 4115-00021
Excellent CIA prepared map to 1:1,760,000 scale.

Map of Saigon - *Khai-Tri Bookshop*, 62 Le-Loi Street, Saigon March 1966
Useful reference for the correct spelling of Saigon's street names; also kept one from getting lost in Saigon in 1968.

United States Air Force Serials 1946 To 1977, Edited by Peter A. Danby - *Merseyside Aviation Society Ltd.*, June 1977
In spite of the title, this is THE source on U.S. Army aviation serial numbers for aircraft used in Vietnam during the years 1954 through 1973.

UNIT HISTORIES

Unit histories were prepared by each unit in Vietnam, efforts being made in the mid sixties to standardize the document format. Often the unit history would include a great deal of information from the unit's previous years; an illustration of the unit patch, etc. and so is included here.

Command Airplane Company Unit History - Extract dated 1970

18th Aviation Company (Utility Airplane Company) 1 January - 31 December 1968

57th Transportation Company (Light Helicopter) 1 January 1963 - 25 June 1963
by 1LT Norman L. St Peter, Unit Historian and 1LT David E. Prewitt, Asst. Unit Historian

121st Aviation Company (Airmobile Light) 1 January - 31 December 1965

125th Air Traffic Company (Air Traffic Control) 1966

145th Combat Aviation Battalion 1 July 1965 - 30 November 1966

Rocket (UTT Helicopter Company) October 1962 - January 1964

SCRAPBOOKS

The 57th Transportation Company, Light Helicopter; The 98th Transportation Detachment (CHFM) and The CH-21C Shawnee compiled by Col. Emmett F. Knight, USA Ret. - Boeing Vertol Company 1987.
Limited edition scrapbook on the 57th, the 98th and H-21's put together, with obvious affection, by someone who knows.

UNPUBLISHED MANUSCRIPTS, DOCUMENTS, INTERNET POSTINGS AND MISCELLANEOUS MATERIALS

List of "FIRSTS" for the 57th Transportation Company (Lt Hel) (H-21) dated 10 August 1962.

Extract from unpublished narrative, dated August 1995, of Col. Bernard B. Quedens, USA Ret. former Operations Officer of the 120th Aviation Company (AML) 1963 - 1964.

Songs of the 57th Transportation Company (Light Helicopter) and its successor unit, the 120th Aviation Company (Airmobile Light) 1963-1964. Courtesy of David E. Prewitt.

3rd Radio Research Unit History (as well as much more on USASA aviation units) by Dennis Buley available as http://www.jnpcs.com/dbuley/usunitsviet.htm.

Army Aviation Patches and Distinctive Insignia. One of the most, if not the most, comprehensive collections of Army Aviation patches of units in Vietnam in existence has been gathered by aviation historian and collector, Chris Miller, of Cobra Company. All other unit patches are credited to the individual or organization who supplied the patch or, if not credited, are from the author's own collection. Most distinctive insignia and "beercans" are from the author's own collection.

PHOTOGRAPHS

All photographs loaned by Army aviators, former unit members who served in Vietnam or aviation historians are credited to the individual who supplied the photograph. All photographs from the U.S. Army, U. S. Air Force and U. S. Marine Corps courtesy Department of Defense. All other photographs, not credited, are by the author or from the author's collection.

This historically significant photograph of the 45th Transportation Battalion Commander, his Staff and his Company Commanders was taken in mid January 1963.

Lieutenant Colonel Robert L. Hoffman, Commanding Officer of the 45th Transportation Battalion (Transportation Aircraft) held a meeting at Tan Son Nhut for his Company Commanders, 14 January 1963. This historic photograph was taken at that meeting. Seated in the front row, left to right, Major Richard D. Kisling, CO 8th Transportation Company (Lt Hel); Major Darwin D. Beauchamp, CO 57th Transportation Company (Lt Hel); Lieutenant Colonel Hoffman; Major Joseph F. Henderson, CO 33rd Transportation Company (Lt Hel); Major George W. Aldridge, Jr., CO 81st Transportation Company (Lt Hel). Standing at back are Major Ivan L. Slavich, CO UTT Helicopter Company; Major Paul R. Ewing, CO 93rd Transportation Company (Lt Hel); Major Morgan H. Mathews, XO 45th Trans. Bn.; Captain Lewis S. Reed, S-4 45th Trans. Bn.; Captain James M. Archuleta, S-3 45th Trans. Bn. US Army Photograph via James L. Bodkin, 145th CAB (Vietnam) Association

IN THE NEXT VOLUME

A De Havilland CV-7A Buffalo, s/n 63-13689, one of two aircraft under test in Vietnam by the U.S. Army Aviation Test Board, photographed in II Corps in December 1965.

Photograph courtesy of Raymond A. Young, III

CV-7 Buffalo Project patch.

Courtesy of Chris Miller

ARMY AVIATION IN VIETNAM 1963-1966

Volume Two picks up where the previous volume left off, with the introduction to combat of newly constituted aviation companies equipped with the Bell UH-1B Huey. Meanwhile, soldiering on until June 1964, the CH-21 Shawnee was finally phased out of the inventory as Army aviation continued its build-up in Vietnam.

During this time frame new aviation battalions, and aviation groups, arrived to assume command of the growing number of aviation companies. Airmobile tactics such as 'Eagle Flights' and large-scale combat assaults were increasingly being perfected. Transportation maintenance companies provided speedy recovery of downed aircraft and helped keep the birds in the air. Fixed wing Otter and Caribou units began to provide a true, dedicated air transport capability to the Army and won a special place in the hearts of US Special Forces. The Army's Bird Dog and Mohawk units were integrated into the reconnaissance effort and became the eyes of the ground commanders.

The enemy had not been idle and had standardized on the AK-47 assault rifle, introduced their first division sized force to combat, and were poised to 'cut the country in half'. The deteriorating situation on the ground, South Vietnamese military reverses on the battlefield, terrorist attacks on the Binks Hotel in Saigon and enlisted billets in Qui Nhon and major enemy attacks against US air bases, led to the introduction of US ground troops in 1965.

The brief, unofficial histories of over 40 Army aviation units will be covered in Volume II. Each unit will be covered with their dates of service in Vietnam, nicknames, brief history, illustrations of their unit patch(es), and photographs of unit insignia, camouflage and markings on aircraft. In addition, Volume II contains specially commissioned aircraft color profiles, uniform artwork and much more.

If you should have any photographs, unit histories, information or memorabilia relating to Army aviation in Vietnam 1961-1973 that will add to this series, please contact the author through The Huey Company, Inc., P.O. Box 625, Ramsey, New Jersey 07446-0625. Especially needed are brigade aviation section aircraft photographs. All material and photographs will be copied, and the originals returned to the sender. Full credit will be given for any photographs or material used and will help to make this series the definitive story of Army aviation in Vietnam.